MW01040684

Christoph Jaroschek

Design of Injection Molded Plastic Parts

HANSER
Hanser Publishers, Munich

The Author: *Christoph Jaroschek*, FH Bielefeld, University of Applied Sciences, Faculty of Engineering and Mathematics, Interaktion 1, 33619 Bielefeld

Distributed by:
Carl Hanser Verlag
Postfach 86 04 20, 81631 Munich, Germany
Fax: +49 (89) 98 48 09
www.hanserpublications.com
www.hanser-fachbuch.de

Editor: Mark Smith
Production Management: Cornelia Speckmaier
Cover concept: Marc Müller-Bremer, www.rebranding.de, Munich
Cover design: Max Kostopoulos
Typesetting: Eberl & Koesel Studio, Altusried-Krugzell, Germany
Printed and bound by Druckerei Hubert & Co. GmbH und Co. KG BuchPartner, Göttingen
Printed in Germany

ISBN: 978-1-56990-893-8
E-Book ISBN: 978-1-56990-894-5

The Author

Prof. Dr. Christoph Jaroschek studied mechanical engineering, and was then Head of Application Technology and Process Development at a well-known manufacturer of injection molding machines for 11 years. Since 1998 he has held the position of Professor in Plastics Processing, in the area of Engineering Science and Mathematics, at the Bielefeld University of Applied Sciences, Germany.

Preface

Many designers are nervous when the material requirement for a development task is a plastic. One reason also lies in the education system or the level of knowledge of the instructors and professors. Until about 1990, the world consumption of plastics was still smaller than the consumption of steels (Figure 1). The volume, not the weight, is used as a benchmark here. If one compares the two groups of materials, volume is a suitable parameter regarding a design, as it is not the weight but the size (i.e. the volume) that is important, especially in the case of plastic components.

Demand for plastics is growing faster than the increase in knowledge among design engineers

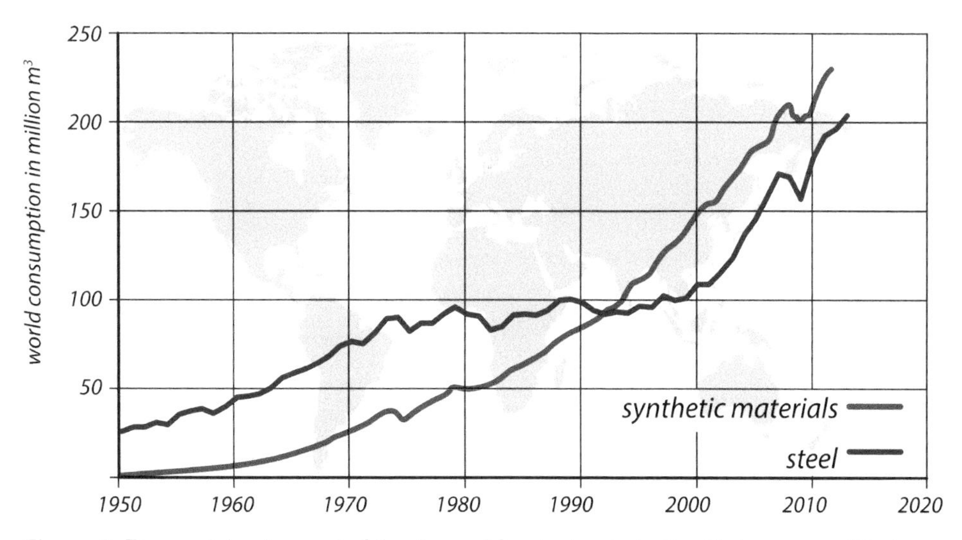

Figure 1 Temporal development of the demand for steel and plastics [Data sources: Word steel assoc., PlasticsEurope Deutschland e. V.]

Before 1990, part design was mainly for metal materials and the importance of plastics was still limited in the training of design engineers. Today, the picture has changed significantly, but it still takes time for training to adapt accordingly.

Dimensioning and calculation

Many books about plastic design have been written by proven experts in the field of plastics themselves. In many cases, the focus is on the calculation or dimensioning of components, rather than on their design. This is where this book comes in.

Injection moldable design

Since most plastic components only must withstand low loads, the actual design of the component is more important than the mechanical design in many applications. It is important to know that most plastic components are injection molded. Therefore, a designer should know first and foremost what this means for the design. This book focuses on the area of injection molds. The designer should be aware that his design specifications must ultimately be implemented with an injection mold.

In this book, the focus is on the field of injection molds. The designer should be aware that his design specifications must ultimately be implemented with an injection molding tool.

Plastics in the following are synonymous with thermoplastics

Due to the focus on injection molded parts, thermoplastics are mainly treated here. These are plastics that melt at higher temperatures. For simplicity's sake, the term plastic is therefore used synonymously for thermoplastics in the following, unless otherwise noted.

Literature

In many places, the content of this book is perhaps a little concise. The book is initially intended to show the important relationships so that the designer understands why the design of injection molded parts must be different from that of metal parts. Many specific details have deliberately not been formulated. This applies, for example, to information on draft angles or radii. Plastics of even one grade (e.g. PP) are available in an almost unmanageable variety with regard to mechanical properties. The properties mentioned here concern, among other things, the mechanical stability under load. Here, a designer should not rely on recommendations from tables but rather consider what effects a too small/large radius, for example, will have for a component. This book is therefore particularly focused on understanding, so that the concrete specifications can be sensibly selected for the respective application.

The compilation of the necessary knowledge for the designer in this book draws on existing literature.

- *Process knowledge*
  - W. Michaeli, H. Greif, G. Kretzschmar, F. Ehrig, *Training in Injection Molding,* Hanser
  - G. Pötsch, W. Michaeli, *Injection Molding,* Hanser
  - T. A. Osswald, L.-S. Turng, P. Gramann, *Injection Molding Handbook,* Hanser
  - S. Kulkarni, *Robust Process Development and Scientific Molding,* Hanser

- *Mold engineering*
  - G. Menges, W. Michaeli, P. Mohren, *How to Make Injection Molds,* Hanser
  - F. Johannaber, *Injection Molding Machines,* Hanser
- *Design*
  - H. Rees, *Understanding Injection Mold Design,* Hanser
  - B. Catoen, H. Rees, *Injection Mold Design Handbook,* Hanser
  - R. A. Malloy, *Plastic Part Design for Injection Molding,* Hanser
- *Material knowledge*
  - G. W. Ehrenstein, *Polymeric Materials,* Hanser
  - T. A. Osswald, G. Menges, *Materials Science of Polymers for Engineers,* Hanser

Contents

1 Plastic Parts

This chapter compares the special features of plastic parts with alternatives made of metal or other materials. There are design rules that are directly justified by the manufacturing process. The information provided here is intended to give the designer a rough overview.

1.1 General Information

Difference between metal and plastic parts

Injection molded components differ from their metal counterparts in interesting ways.

- Plastic parts have a different shape for the same function.
- Often a conventional assembly can be realized in one plastic part, i. e. many functions can be implemented directly in a single component.

For example, consider compressor bars made of metal, plastic, and material combinations (Figure 1.1). First, it is important that the requirements are met. The question as to which material is better or worse is not possible until clear evaluation criteria have been established.

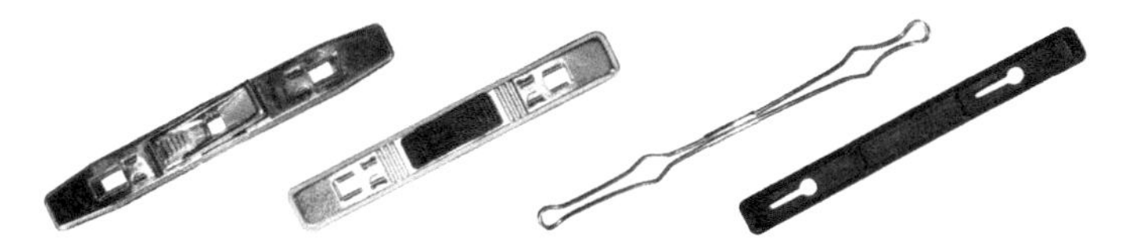

Figure 1.1 Metal and plastic compressor bars for ring binders

General requirements

In any event, the requirements for the compressor bars are:

- Function: Clamping force
- Economy (manufacturing costs).

The clamping force is generated by the deformation of a wire in the elastic range in the case of the metal variants and by the deformation of the plastic in the all-plastic variant. Due to the considerably lower modulus of elasticity of plastic, the plastic variant is only suitable for small forces and should not be used for very thick ring binders.

Manufacturing costs of injection molded parts are only favorable for large production runs

Manufacturing costs consist of the costs of material, production equipment (machine and mold) and labor. Roughly speaking, material costs constitute half of the manufacturing costs. Material costs range from 2 to 4 $/kg. In the all-plastic variant, the costs are very low because the product is created in a single process step. Although the machine and tooling costs are very high, if the expected number of pieces exceeds the limit of about 10,000 the tooling costs per part are low. And if many injection molded parts can be produced per hour with one machine, the machine costs per part are also low.

Functional integration leads to simpler production

The metal compressor bars consist of several elements that must be joined together. Basically, the fewer process steps that are necessary, the lower is the risk of failure in production. This should also be considered when compiling manufacturing costs.

1.1.1 Comparison of Designs (Conventional vs. Plastic)

Rethinking the design when using plastics

The use of plastics requires a fundamental design rethink. The example of a clothespin shows that the older product made of wood is cheaper than a similar plastic clamp (Figure 1.2). Both variants consist of two clamp elements that are pressed together by a metal spring. The wooden clamp can be cut very quickly from a profile-milled board. The corresponding plastic clamp is more expensive to manufacture and has inferior properties, because it can become brittle and break due to weathering.

A well-designed plastic clamp will consist of only one element, and that eliminates the need for assembly. In principle, plastic components can incorporate many functions. This is referred to as functional integration.

Cast structures can have very freely formed surfaces

A plastic component can have a very complex design if it is manufactured by injection molding. Due to the molding process used for production, the design of a plastic component can feature any type of free-form surface. In conventional components, the individual parts are predominantly milled and turned from the solid, with the result that simple shapes predominate here.

Considering a comparison to conventional products, the following generalization can be made:

Conventional components often consist of various individual parts that form an assembly. By contrast, good plastic components often consist of a single part (Figure 1.3).

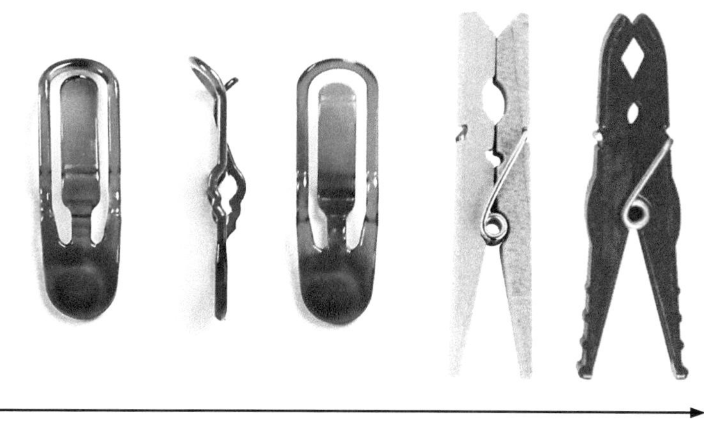

Figure 1.2 Manufacturing costs of clothespins of different designs

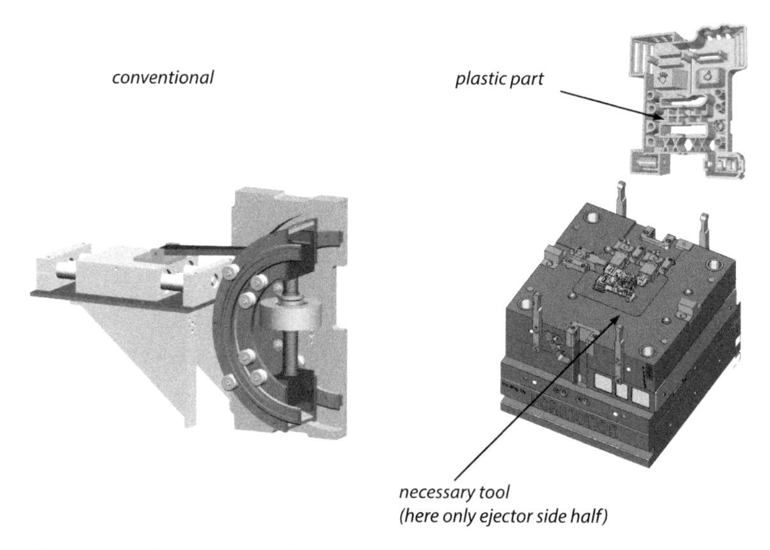

Figure 1.3 Comparison of a conventional assembly consisting of different individual parts and a plastic component, along with the mold required for production [image source: Ziebart/FH-Bielefeld, Ritter/HS-Reutlingen]

Good designs in plastic consider the feasibility of using injection molds

During the development of a plastic component, consideration must be given to the mold at the design stage, because it limits the design freedom to a certain extent. In any event, the designer of an injection molded part should be aware of the possibilities afforded by mold technology, because slight changes in the shape of a plastic component can have a very large effect on the cost of a mold. Molds consist of many individual parts and are in turn very complex assemblies. The molds must perform different tasks (Figure 1.4). The actual mold cavity has to be filled with melt and the heat of the melt needs to be dissipated (cooling) so that the plastic part will become solid and stable and can be demolded via an ejector system.

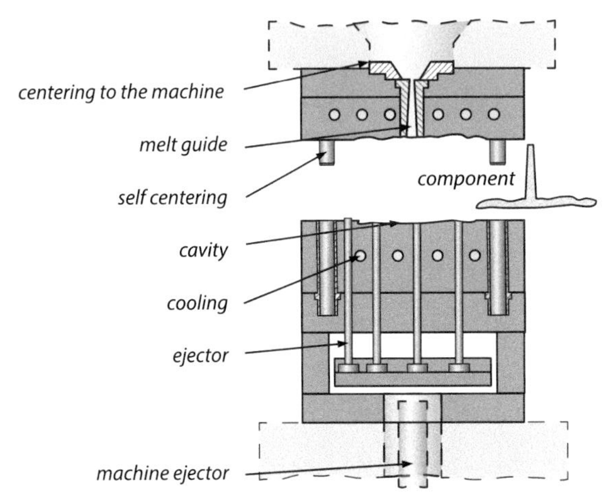

Figure 1.4
Design and functions of a simple injection mold

Demonstration mold shows effect of good component design on the mold

For demonstration purposes, the "Polyman" plastic component shown in Figure 1.3 is poorly designed on the left side and well-designed on the right side. This assessment of the design relates to the mold implementation. For the various lateral openings, three sliders are required on the poorly designed side to demold the undercuts (Figure 1.5). With a few minor changes to the shape, the well-designed side can dispense with sliders completely. This makes the mold less expensive and less susceptible to faults during production or requires less maintenance.

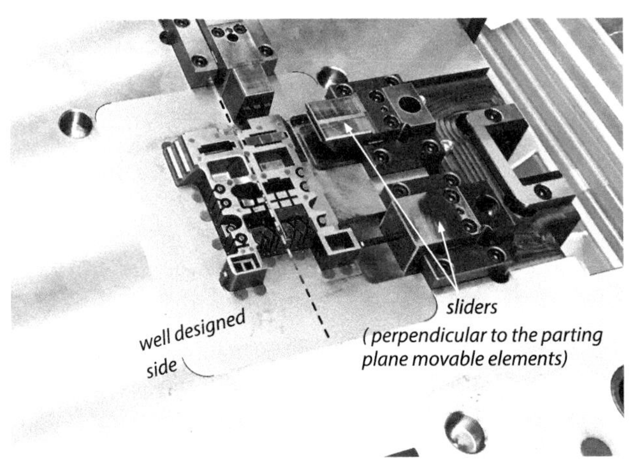

Figure 1.5
Ejector side for the Polyman demonstration part [image source: Ritter/HS-Reutlingen]

1.1.2 Special Features of Plastics

The biggest advantage of plastics is their low melting point

The most important property is the melting temperature of plastics, which is only about 1/10 that of metal (Figure 1.6). This makes it possible to cast plastics in steel molds of very complex shape. The precision of the steel molds can be transferred to the plastic component largely without the need for reworking and can be repeated almost as often as required. However, the complexity and expense of such molds render this production barely suitable for small production quantities. Plastic parts manufacture is thus almost always a mass production process.

A distinction needs to be made between melt temperature and transition temperature. In processing, the melt temperature is always much higher than the transition temperature from the solid to the melt state. Strictly speaking, only semi-crystalline plastics can melt, because melting entails the liquefaction of crystalline areas. Amorphous plastics, therefore, merely soften. This may not become clear until Chapter 5, where specific material properties and characteristic temperatures are discussed.

Temperature of the plastic melt

1.1.2.1 Comparison of the Properties of Plastics and Metals

Further comparison with metals reveals major differences in properties. Thus, specific applications may only be feasible in one of the two materials.

Mechanical properties of plastics are not as good as those of metals

Table 1.1 Comparison of Metals and Plastics

Property	Metal	Plastics
Young's modulus	high	low
Tensile strength and yield strength	high	moderate
Density/weight	high	low
Young's modulus	no	possible

- The modulus of elasticity of metals and especially steels is approx. 1000 times higher than that of plastics. Applications subject to high load requirements are therefore largely limited to metals. Plastic components would deform too much in such cases.
- Metals are stronger than plastics. The issue here is that of component failure. This can be both a fracture and an unacceptable permanent deformation.
- Young's modulus and the strength of plastics are strongly dependent on temperature. For applications involving high temperatures, which can be as low as 50 °C, particular care must be exercised in the choice of material subject to long-term loads.

Young's modulus for plastics is temperature-dependent and therefore not constant

- The density of plastics is only approx. 1/7 that of steel. Applications that require a certain weight (e.g. pendulums for clocks or curtain weights) cannot easily be made in plastic.
- Some plastics are transparent.

A close comparison of different materials reveals further advantages and disadvantages.

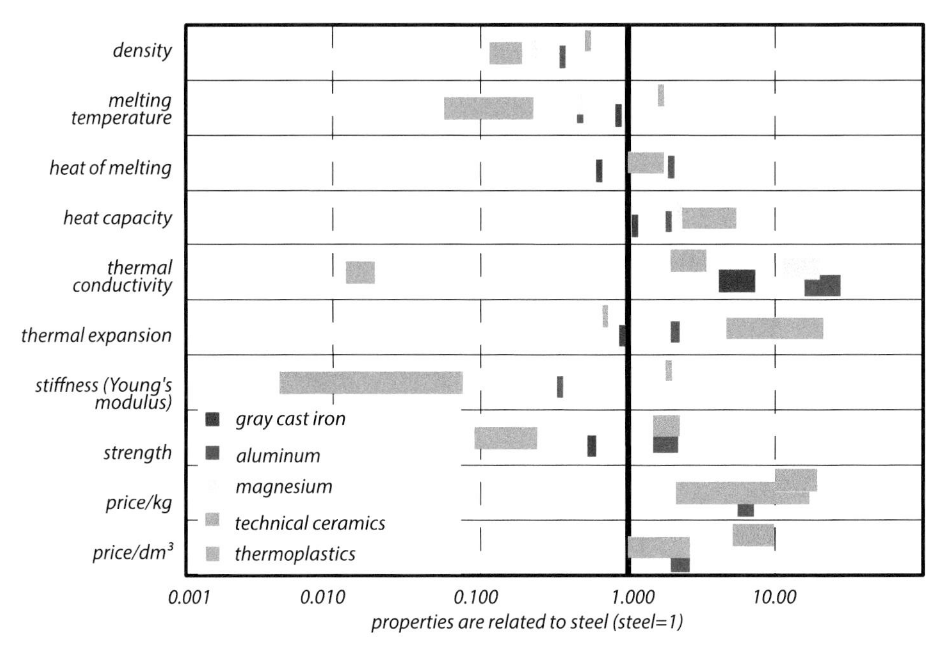

Figure 1.6 Comparison of thermoplastics with steel [source: WAK-Kunststofftechnik]

With regard to production, the low thermal conductivity of plastics makes it difficult initially to dissipate the heat of the melt from inside the component to the mold. The thicker a component is, the longer the cooling process will take. For this reason, plastic parts are thin-walled wherever possible. Jumps in wall thickness are unfavorable.

Low thermal conductivity enables precision injection molding of fine structures

The low thermal conductivity, however, also makes it possible to fill long, thin flow paths in a controlled manner. Plastic components can thus be considerably finer structured than cast metal components.

Plastics are usually more expensive per kg than metals

It is often assumed that plastics are inexpensive, but this is not the case. Especially those plastics that are intended for use at elevated temperatures can cost more than $10 per kilogram. When expressed in terms of weight, the outcome is the specific raw material price, which is given in $/kg. This is comparable to that of metals.

1.1.2.2 Special Mechanical Behavior

Metals have a definite failure limit (yield strength)

Metals are atomic in structure, i.e. they are composed of individual atoms that form crystals in regular repetition during cooling. When a load is applied, the atoms move slightly away from each other, returning to their original state after the load is removed. This elastic behavior is linear, i.e. the deformation increases in proportion to the load. Above a load limit R_p, entire atomic layers shift; when the load is removed, the deformation remains (Figure 1.7).

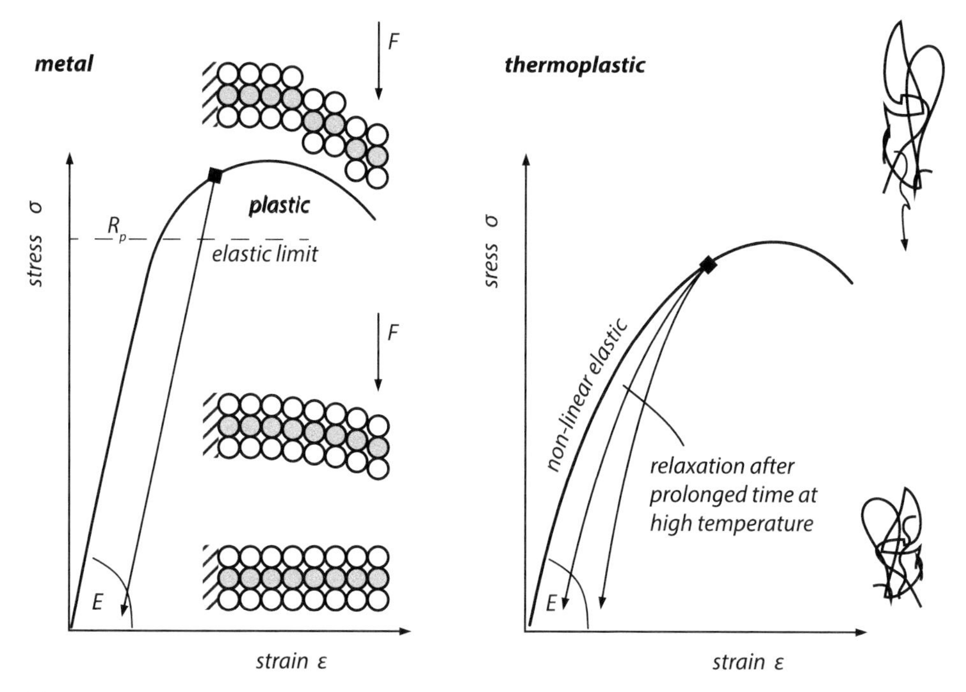

Figure 1.7 Elastic and plastic deformation of metals and plastics at different load levels

Mechanical failure of plastics is time- and temperature-dependent (viscous)

Plastics have a molecular structure, which can be imagined as a collection of tangled spaghetti. The entire tangle of molecules is initially elastic under load. At low temperatures, the molecular chains "stick" to each other; at higher temperatures, they are able to slide off each other. The deformation behavior is not linearly elastic, i.e. at high loads, the deformation becomes increasingly greater. The extent to which a load allows complete recovery of deformation depends strongly on its duration. A brief load is often elastic, while a load applied for a long time causes irreversible viscous deformation.

Difference between plastic and viscous deformation

For accuracy, the deformation of plastics should not be called plastic, but viscous. Plastic deformation always describes an irreversible deformation; in metals, this occurs through the sliding of planes of regularly arranged atomic layers. Plastics, however, consist of tangled long molecules which, under load, can deform as a whole bundle and can partially recover their deformation. Viscous deformation is said to occur when the permanent deformation that cannot be recovered is time-dependent.

Thermoplastics can soften or flow at higher temperatures

The different plastics each have a specific molecular structure. There are two groups (Figure 1.8): crosslinked and non-crosslinked. Thermosets and elastomers belong to the crosslinked plastics. During processing, chemical bonds form between the molecular chains, preventing them from melting even at high temperatures. The following pages deal almost exclusively with non-crosslinked thermoplastics. A distinction is made between amorphous and semi-crystalline thermoplastics. Amorphous means that the molecular chains are entangled totally irregularly. In the case of semi-crystalline plastics, it is possible for some of the molecules to form regular arrangements in crystal form during the cooling process.

non-crosslinked plastics		*cross-linked plastics*	
amorphous	*semi-crystalline*		
loosely entangled molecules	*entangled molecules and molecules linked by crystals*	*molecules cross-linked,*	*molecules cross-linked, soft intermediate segments,*
$T_{operation} < T_{glass}$ *predominantly brittle* *at* $T > T_{glass}$ *plastic formable*	$T_{operation} < T_{melt}$ *at* $T < T_{glass}$ *brittle/elastic* *at* $T > T_{glass}$ *tough* *at* $T > T_{melt}$ *formable*	$T_{operation} < T_{degradation}$ *predominantly brittle* *non-meltable*	$T_{operation} < T_{degradation}$ *at* $T < T_{glass}$ *brittle/elastic* *at* $T > T_{glass}$ *tough* *non-meltable*
thermoplastics		*thermosets*	*elastomers*

Figure 1.8 Structure and application range of plastics

Below the glass transition temperature, plastics are mostly brittle and hard

All plastics have a characteristic glass transition temperature. Below this temperature, the behavior is largely glassy, i.e. brittle. The individual molecules of the plastic are virtually frozen and cannot be displaced against each other. Permanent deformation is not possible; the plastic behaves elastically. Above the glass transition temperature, the plastic softens increasingly, turning from tough to soft. Under load, permanent deformation occurs to an extent that depends on the temperature and the duration of the load.

Above the melting temperature, crystals of semi-crystalline plastics melt

Semi-crystalline plastics have an additional characteristic melting temperature above which the crystals melt. While amorphous plastics can only be used below the glass transition temperature because, above it, they undergo noticeable viscous deformation under load, semi-crystalline plastics can be used up to the melting temperature range. The crystals hold the molecules together. Compared to crosslinked plastics, the crystals act as physical crosslinking points.

The load-bearing capacity of plastics is temperature-dependent

Mechanical behavior is strongly dependent on temperature, especially in the case of thermoplastics (Figure 1.9). At very low temperatures, all plastics are predominantly brittle and break without warning like hardened steels (behavior 1, also characteristic of thermosets). In the glass transition temperature range, slight necking occurs shortly before fracture (behavior 2); unlike metals, the maximum here is called the yield stress. Semi-crystalline plastics can deform extensively in the range between the glass transition temperature and the melting temperature. Very characteristic here is the shoulder-neck deformation (behavior 3). For actual applications, however, this behavior is of no further significance, because plastics should only be loaded up to their yield stress.

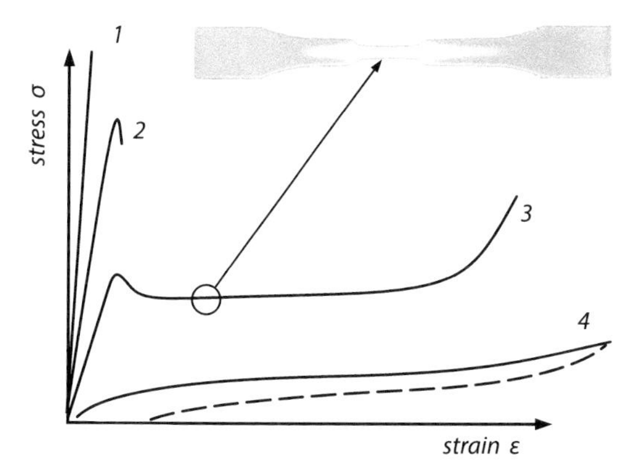

Figure 1.9 Deformation behavior of plastics under load
1: Thermoset or amorphous thermoplastic
2: Thermoplastic in the range of the glass transition temperature
3: Semi-crystalline thermoplastic above the glass transition temperature
4: Elastomer

It is interesting to note that semi-crystalline plastics in particular exhibit behaviors 1 to 3, depending on the operating temperature. Note that here the slope of the stress-strain curve becomes increasingly flat with increase in temperature, i.e. Young's modulus becomes smaller with increase in temperature.

Plastics for high-temperature use are more expensive

The molecular structure influences the glass transition temperature and, in the case of semi-crystalline plastics, the melting temperature. The heat deflection temperature characterizes the temperature up to which a plastic component still largely preserves its shape under load. This property is closely related to the raw material price (Figure 1.10).

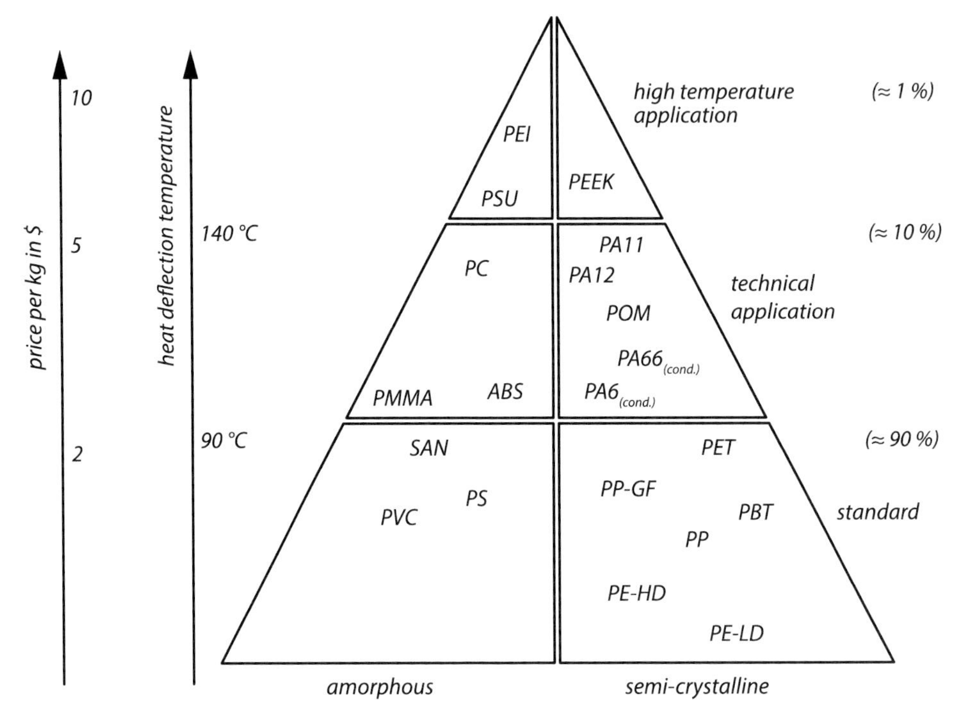

Figure 1.10 Important amorphous and semi-crystalline thermoplastics

The most important plastics are PP, PE, PVC

Even though there are very many different plastics, only a very few grades are used (Figure 1.11), with PP and PE playing a major role. This is because the properties of plastics can be modified within wide limits by means of fillers. This largely applies to stiffness (Young's modulus), which can be increased with glass fibers, for example.

Engineering plastics of high heat resistance are mainly used in the automotive sector.

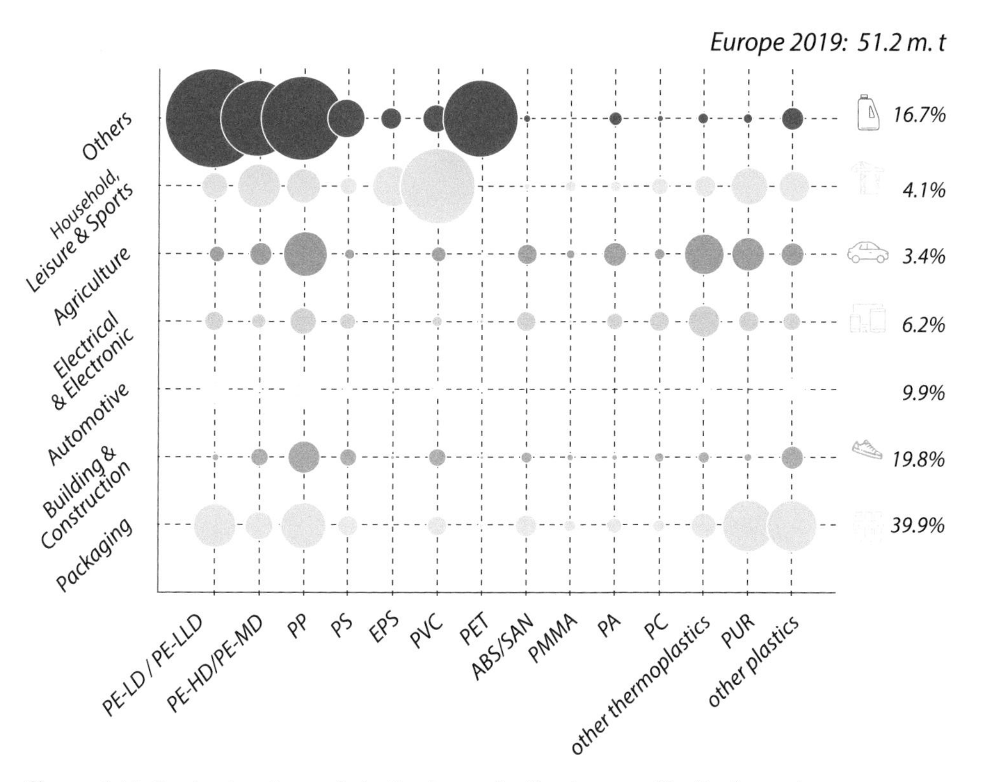

Figure 1.11 Predominantly used plastics by application [source: PlasticsEurope]

1.1.3 Reasons for Using Plastics

Plastics are chosen for a variety of reasons:

- Cost-effective reproduction
- Functional integration
- Material substitution
- Special properties
- Design
- Weight savings

Mass production requires cost-effective, repeatable production

Cost-effective reproduction means mass production. The cost of a single injection-molded component is very high because it cannot be produced without a suitable mold. Injection molds can be very expensive, with the price also depending on the complexity of the component to be produced. Complex components made of plastic are therefore always mass-produced items because the mold costs are spread over many components. Injection molding becomes an interesting proposition from a batch size of 10,000. Batch sizes below this limit, if produced at all, are more likely for geometrically simple components.

With one manufacturing process, several functions can be realized at the same time

Functional integration means that a component directly fulfills several functions. Because plastic components are predominantly molded, combinations of different functions are usually easy to implement. One such example is a dishwashing brush (Figure 1.12). Changing the material from the original wood, wire and bristles to plastic and bristles makes it possible to add a scraper function for removing stubborn dirt residues without much additional outlay. The process can be extended to add a suction foot, along with a soft component. These additional functions make the products competitive and, in some cases, less expensive than products made of conventional materials. The designer needs to be aware that molded designs offer many degrees of freedom. It is very often worthwhile to consider incorporating additional functions.

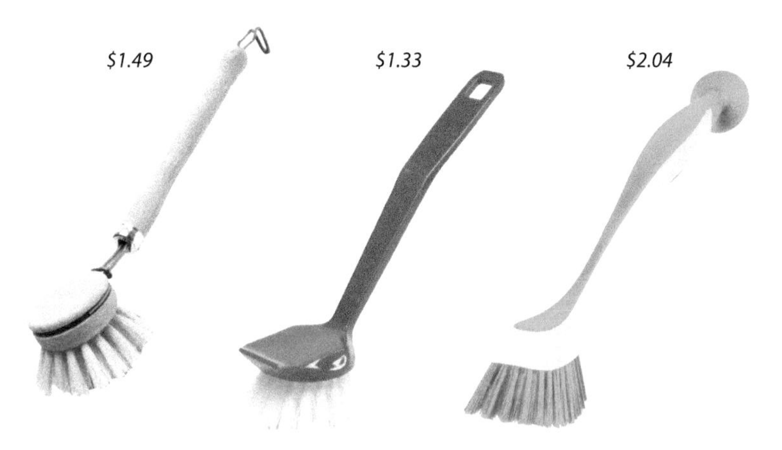

Figure 1.12 Examples of functional integration, in which the original component functions are extended by simple means

Plastics instead of metals for weight and cost savings

Material substitution means replacing one material with another. In contrast to the example of functional integration, the function and component design are largely retained in this case. Reasons for replacing a material with plastic could be:

- Weight reduction
- Reduction in manufacturing costs

A change of material from metal to plastic requires a redesign

However, switching to plastic inevitably results in geometrical changes. Material accumulations have to be eliminated, and additional stiffeners may be necessary (Figure 1.13). Plastic components often have a better surface quality for visible parts and are corrosion resistant without the need for post-treatment, such as anodizing, coating, or painting.

Figure 1.13 Example of material substitution: here, a switch from aluminum to polyamide

Plastics have various **special properties** that render them useful. In many cases, these properties can significantly improve the original functionality. The example of a lemon squeezer clearly shows that, in addition to the original squeezing function, an elastic silicone provides extra functions, such as storage (Figure 1.14).

Very special properties for very special applications

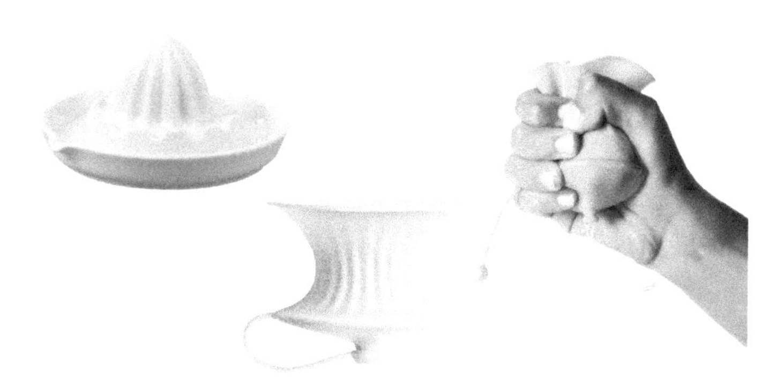

Figure 1.14 Extended functionality and design possibilities due to the very high flexibility of silicone as exemplified by a lemon squeezer [image source: Jähn, Hachenburg]

The **special properties** of a select number of plastics include electrical non-conductivity, acid resistance, and corrosion resistance. In this case, the chosen plastic must possess the relevant property profile.

Design relates to the actual shaping. The example of the door wedge illustrates the possibilities offered by a change of material. In addition to the pure design aspect, the properties also come into play (see Chapter 5, "Material Selection"). The plastics employed also differ from wood in their coefficient of friction and hardness,

Plastic components can have very interesting shapes

with the result that their original function as door-stopper is enhanced (Figure 1.15).

Figure 1.15 Plastics molding processes offer extended design possibilities over conventional processes involving the sawing and ablation of wood

■ 1.2 Design Rules

Design of plastic components is about the necessary tooling technology for production

If possible, the designer should always have the manufacturability of the design in mind. Injection-molded parts are mass castings, giving rise to three special characteristics that have a direct effect on optical quality:

- The process requires at least two mold halves for demoldability. The parting edge between the mold halves is almost always visible and tangible. Sliders are often necessary for demolding, which are moved transversely to the opening movement to demold undercuts. These sliders also form parting lines that can be seen and felt.
- One of the mold halves contains the ejector system. The pins, which are cylindrical in the simplest case, stand out on the component surface (Figure 1.16).
- The gating point usually leaves a clearly visible tear-off point

Injection molding inevitably produces surface defects

With an awareness of these self-evident facts, quality losses can often be minimized by relocating the gating point to non-critical areas, allowing the parting line to follow component edges as far as possible, and relocating slider separations to the non-visible side if necessary. Another possibility is to use a suitable surface structure texture that hides the above-mentioned markings.

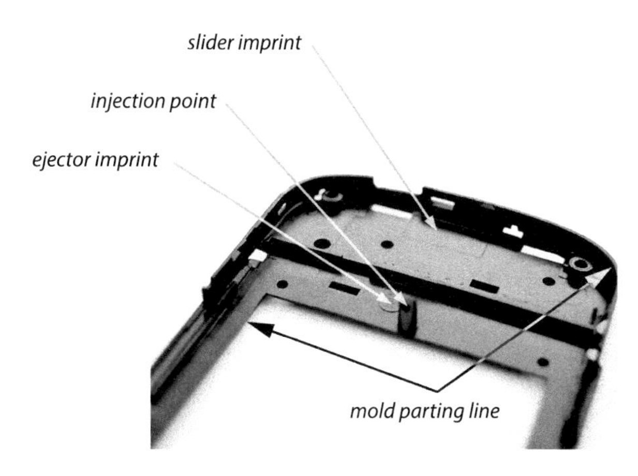

Figure 1.16
Inevitable markings or optical losses on injection molded parts

Separation of the mold halves is visible on the component

In many injection molded components, the parting line also separates the visible side and the non-visible underside. Often enough, however, the parting line is also located directly on a component's surface.

In any event, it is advisable to design the separation directly. For the development of the injection mold, the surface of the CAD part is divided into two surfaces for the two mold halves (nozzle side and ejector side) and a parting surface is added in each case (Figure 1.17). CAD programs feature Boolean operations for forming a solid from the union or intersection of two solids. If the CAD component with the parting surface is placed in a cuboid solid and a subtraction is then performed, the negative impression of the component is obtained directly, in the form that will later be seen in the cavity (Figure 1.18).

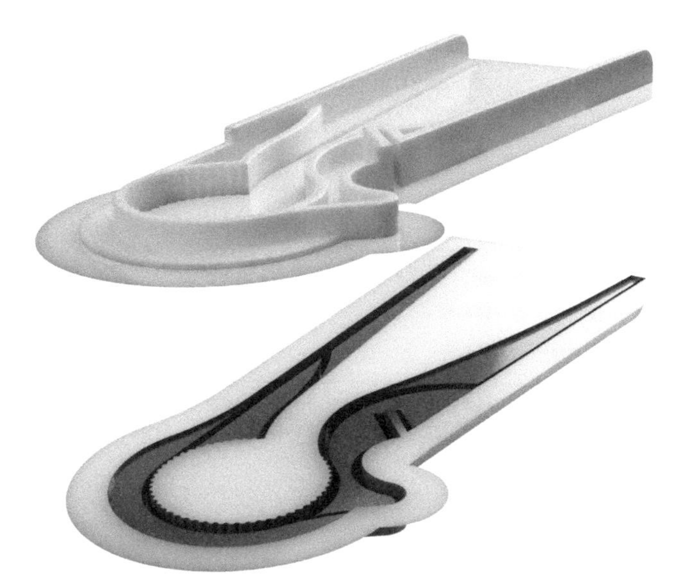

Figure 1.17
Split surface of a component for the core and nozzle sides, with attached parting line [image source: S. Schneiders]

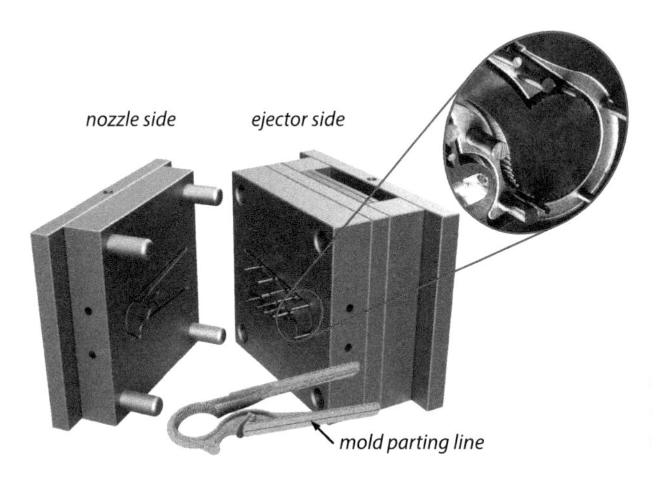

Figure 1.18
Mold for a bottle opener, showing the negative contour of the part surface [image source: S. Schneiders]

1.2.1 Special Design Features of Injection Molded Parts

The injection molding manufacturing process imposes the following special design features on component:

- It must be possible to demold the component.
- The possible flow-path lengths are limited.
- The gate position is always a defect and sometimes a weak point.
- Material accumulations should be avoided.
- Components should be as uniformly thin as possible.
- Ribs, beads, and crowns make components stiffer.
- The dimensions of plastic components change significantly in line with the service temperature; this merits special consideration in the case of assemblies made of plastic and metal sheets.

1.2.1.1 Demoldability

Designer must think in terms of mold halves

It is best for the designer to think in terms of mold halves (Figure 1.19). This does not mean the mold details, but rather the basic structure and arrangement of the mold halves that give the mold its shape.

The component's shape essentially determines the complexity of an injection mold

The designer should give consideration to the position of the component in the mold and the separation of the mold halves. In many cases, doing so will avoid undercuts and transverse holes, which make demolding difficult. Otherwise, sliders or core pulls become necessary, but this makes the mold more expensive and more prone to failure because of the additional moving elements (Figure 1.5).

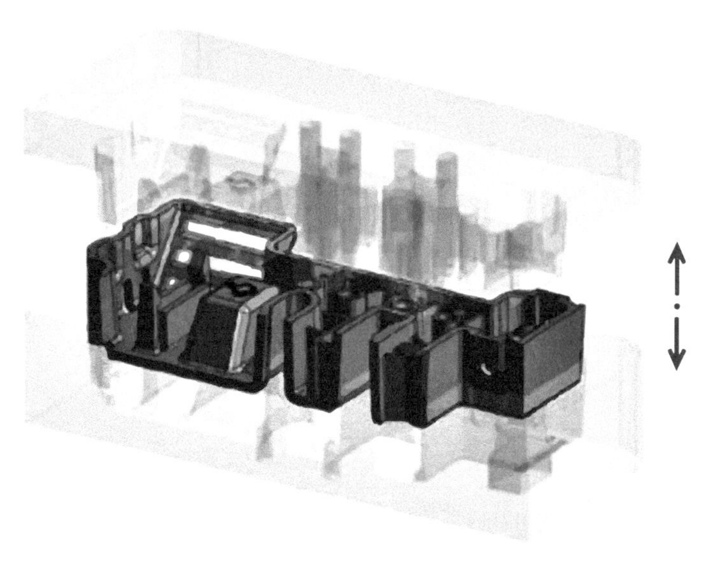

Figure 1.19 Component design always factors in the necessary mold halves [image source: Ritter]

During solidification, the plastic shrinks and if the component is tall enough, it will shrink onto the core (Figure 1.20), which forms the inner shape of the mold cavity. Demolding drafts are therefore very important for ease of demolding because they mean that high forces (breakaway forces) only have to act on the component over a short path in the initial stages of demolding.

Demolding drafts facilitate or enable demolding

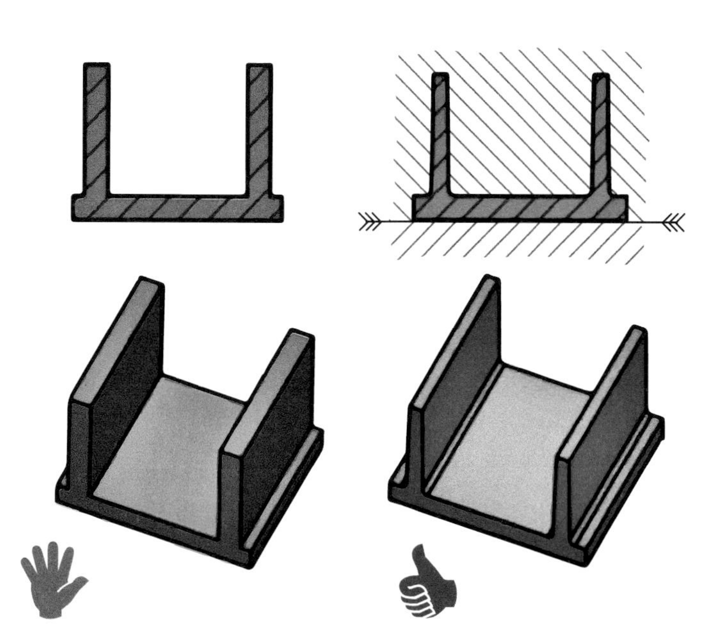

Figure 1.20 Draft angles reduce the demolding forces [image source: Ritter]

Variables influencing the draft angle

The draft angle depends on:

- Shrinkage potential of the material: Semi-crystalline plastics shrink more than amorphous plastics and fillers reduce the shrinkage potential.
- PP and PE have particularly high shrinkage values.
- Demolding height: Tall, slender cores are problematic.
- Roughness depth: This refers to the surface condition of the mold. In general, 1° per 0.025 mm roughness depth can be assumed. Surface depressions of 0.1 mm are quite possible in the case of grained surfaces.

Table 1.2 Draft Angles for Various Plastics

Abbreviation	Draft Angle
PE	0.5°-1°
PP	1°-3°
POM	0.25°-0.5°
PC	0.5°-1.5°
PPO	0.5°-1.5°
PBT	0.5°-1.5°
PA 66	0.125°-0.5°
GF-PA 66	0.25°-1°
ABS	>0.5°

The direction in which a component will shrink is not always clear, and in many cases several cores are located next to each other on one side of the component. Therefore, steep walls have a slope on each of the two sides.

Draft angles on tall, slender cores lead to different wall thicknesses, depending on the mold design

In the case of tall, slender cores, an unfavorable choice of mold parting line can lead to large differences in wall thicknesses, or the component may become too thin or too thick in some areas (Figure 1.21, left and center). If the parting line is placed at half height, the component tapers toward two sides and the difference will be smaller. In the illustration on the left and in the center, the contour of the component would have to be completely machined into the upper mold half; where this is necessary, it can only be done by means of EDM.

The wall thickness is even more uniform if the parting is carried out using the mold core-cavity parting method, whereby the entire inner area of the component is formed by one mold half and the outer area by the second mold half (shown on the right). This has the added advantages of easier manufacturability and easier venting. The option shown on the right of the Figure 1.21 could well be produced by conventional machining; the core and die are each in a separate mold half. This option would produce a uniform wall thickness over the height of the part, with the draft slope running parallel on the outside and inside. Note, however, that in this version the diameter at the top of the sleeve is smaller than at the bottom of the part, because of the necessary draft angle in this case.

With regard to venting, it should be remembered that the air displaced by the inflowing melt can only escape from the mold cavity via the mold parting line in each case, unless it is diverted by means of an additional venting pin, as shown on the right of Figure 1.21, in the upper mold half.

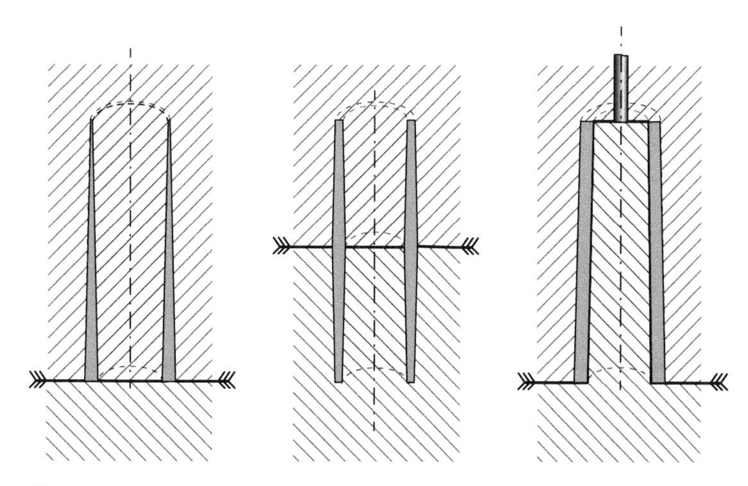

Figure 1.21 Draft angles on tall, slender cores

Design of openings without undercuts

Components often have lateral openings, e.g. for ventilation slots or viewing windows. If the design is not very carefully thought out, this inevitably results in an undercut which can only be removed by an additional transverse movement of a slider. There are various design ways for avoiding undercuts. In Figure 1.22 a cross-section is shown through the component for each design and the mold halves are indicated by hatching. All designs on the right allow the mold halves to be moved perpendicularly to the parting line without colliding with the component.

Similarly, undercuts can be designed without sliders if an opening parallel to the parting can be permitted. In this way, the lower half of the mold can form the underside of the snap-fit hook in Figure 1.23.

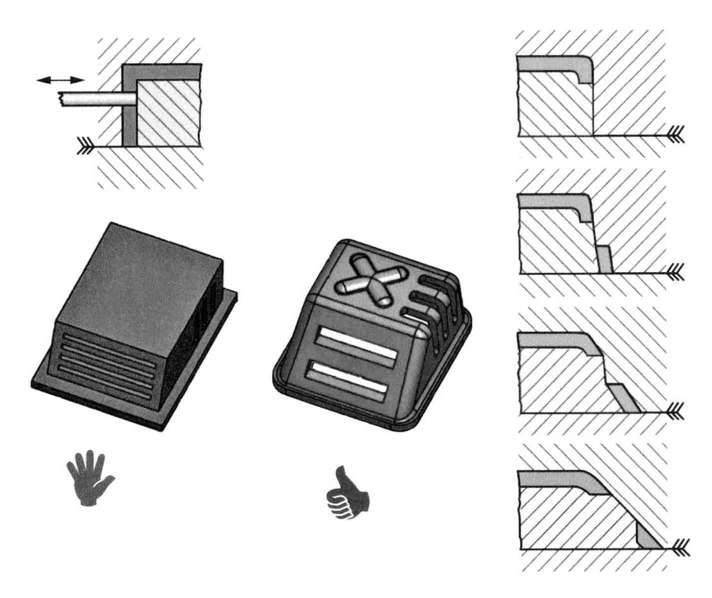

Figure 1.22 Design of openings for, e. g., ventilation slots

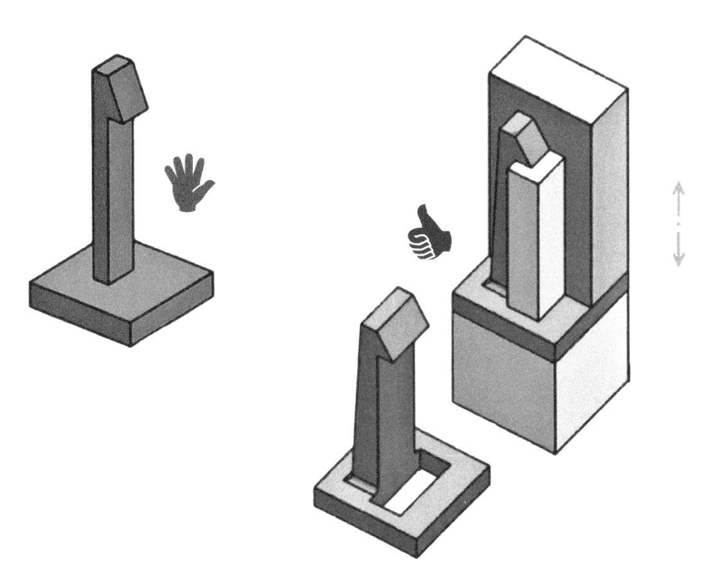

Figure 1.23 Design of a snap-fit hook without the need for a slider [image source: Ritter/HS-Reutlingen]

A combination of the design for snap-fit hooks with opposed alignment is suitable for axle bushings if a precision-fit, completely round bore does not have to be created (Figure 1.24).

Design of axle passages

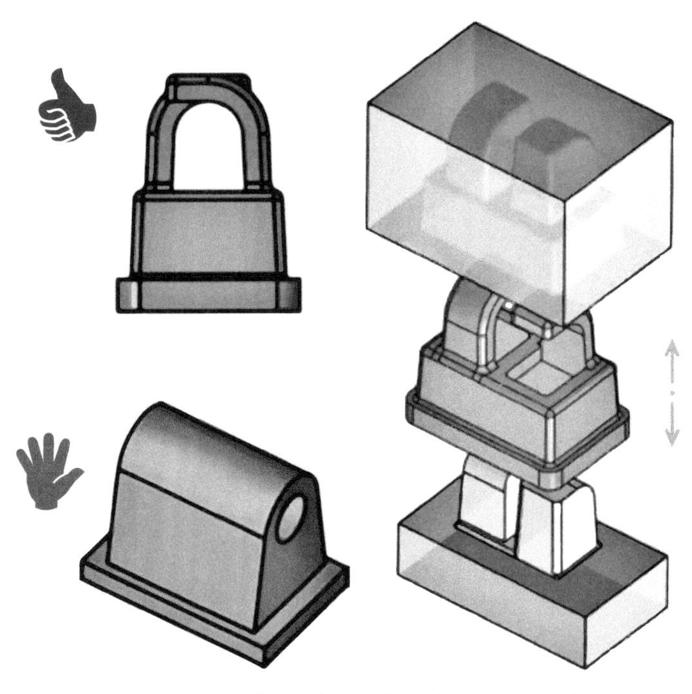

Figure 1.24 G design of an axle bushing without core puller or slider [image source: Ritter, HS-Reutlingen]

Angular elevations for, e.g., screw holes can make the mold unnecessarily expensive. The designer should be ever mindful that a slim, tall component contour always requires a narrow, deep counter-contour in the mold. In the mold, these areas can often only be produced in several parts and by inserting a corresponding contour insert, which can be easily machined from the outside, into a matching recess in the mold (Figure 1.25). The same function can often be achieved with a modified contour, although the round shape shown in the picture requires less outlay in the mold.

Inserts required by the design

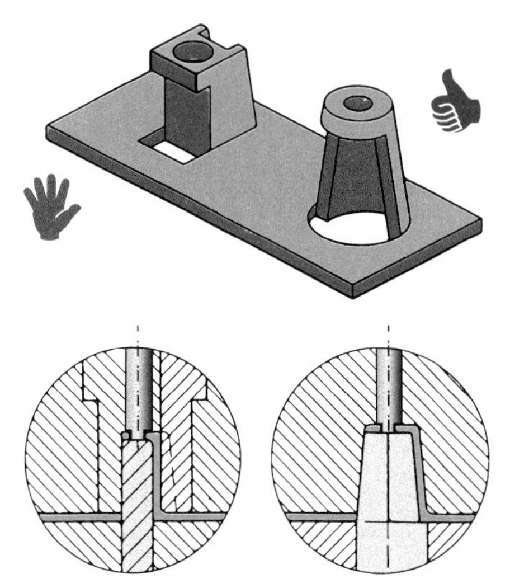

Figure 1.25
Design of "angular" attachments and protrusions [image source: Menges]

Mold inserts are not fundamentally bad, but they increase the complexity and thus the costs. The functionality of a screw boss remains the same with a modified shape, but it is considerably easier to implement in terms of tooling technology. In Figure 1.25, a pin is inserted in the upper half of each mold, which in this case can also be used for venting. Even with very close tolerances between the mold elements, gaps are created through which the air can escape from the cavity area during injection.

1.2.1.2 Flow Path to Wall Thickness Ratio

Injection molded components have a limited flow path to wall thickness ratio

Flow-path lengths are limited by the flowability of the plastic. For normal technical components made with common materials, the flow path to wall thickness ratio is approx. 150:1 (Figure 1.26). If fillers are used, flowability and thus the possible flow-path length decreases.

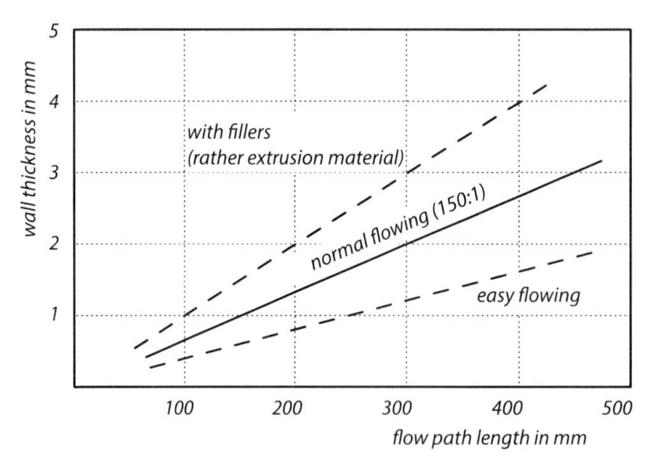

Figure 1.26
Possible flow-path lengths as a function of wall thickness

The flow-path length is not given consideration by many designers. Long flow paths are only possible if several injection points are provided, but these then require hot runners in the mold to keep the melt at the melting temperature between the central melt feed to the mold and the injection points. This inevitably requires greater outlay on mold technology and possibly also leads to surface defects when the melts from the various injection nozzles flow together.

Large components often have to be molded in at several injection points.

1.2.1.3 Sprue Position

The choice of gating point significantly influences the quality of the manufactured component via:

Influence of sprue position on component quality

- The occurrence of weld lines
- The possible formation of air inclusions
- The alignment of fibers

Weld lines occur when two melt fronts meet. This always happens behind cutouts, i.e. when the melt has to flow around mold cores. But melts can also meet in the absence of cutouts if, for example, different wall thicknesses influence the flow of the melt. The melt flows faster in thicker component areas because of the lower pressure needed there. If wall thickness variations exist, it can advance faster in the thicker areas. This results in two flow fronts, because the melt flows from the thicker area into the thinner adjacent area, where it can merge with the flow front of the slower-flowing melt (Figure 1.27).

Weld lines are formed at confluence points, even without cutouts

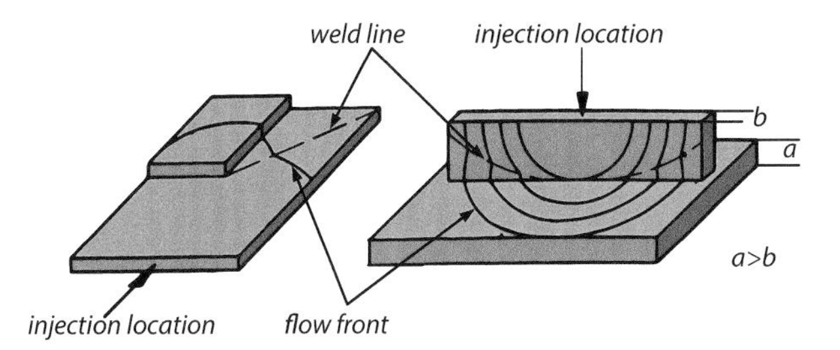

Figure 1.27 Possible weld lines on components depending on component design and injection location

Where weld lines occur is not easy to predict in the case of complex component designs. In such cases, flow simulations are highly reliable. With little effort, these can be used to visualize the effects of different gating locations. Wall thickness changes require somewhat greater outlay because the geometry change often takes place in the CAD system and then a new simulation model (FEM mesh) has to be generated.

Weld lines can be predicted with simulation programs

Common programs for flow simulation also allow several gating locations for a component or also temporal control of the opening at several gating locations (see Chapter 4). This has the advantage of even making possible long flow paths without weld lines.

Injection location influences the properties of fiber-filled plastics

In fiber-reinforced thermoplastics, the injection point itself also significantly influences the orientation of fibers and thus the mechanical, direction-dependent properties as well as direction-dependent shrinkage. For fiber-reinforced materials, the properties are dependent on the flow direction. Estimating the orientation of the fibers and the effect on the mechanical properties is almost impossible without recourse to simulation.

1.2.1.4 Avoiding Material Accumulation, Thin Wall Thickness

The component thickness essentially determines the necessary cooling time

Plastic components are usually thin-walled and should have a uniform wall thickness. Due to the poor thermal conductivity of plastic, the cooling time becomes longer with increase in wall thickness. The cooling time can be estimated from

$$t_{cooling} = s^2 \, \text{s/mm}^2 \quad (1.1)$$

where s is the wall thickness.

Wall thickness reduction through coring

Thick component areas should always be avoided. If necessary, it should be considered whether such areas can be cored out. In that event, a mold core fills the thick area; this is always possible if it is not necessary for both surfaces of a shell-shaped component to have a defined shape (Figure 1.28).

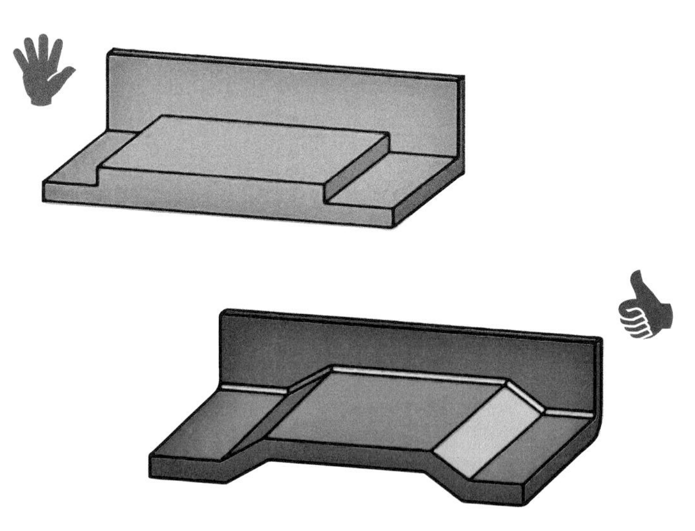

Figure 1.28 Thickness of thick areas can be adjusted by coring

Coring can also be used for cylindrical geometry areas (Figure 1.29). When designing, consideration should always be given to whether the conventional geometry is necessary or whether a more favorable shape for the process might be possible.

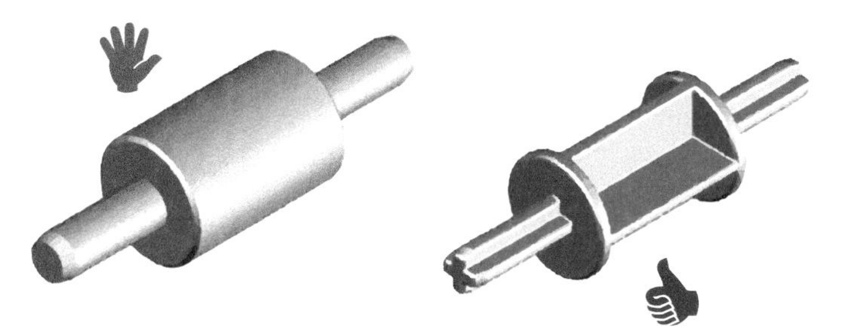

Figure 1.29 Coring leads to uniform wall thicknesses [image source: Protolabs]

Stabilization of screw bosses

Screw bosses are often connected to the side walls for stability reasons. To avoid uneven wall thicknesses, the connections can be made by means of ribbing (Figure 1.30). However, it should be borne in mind that the concavities next to the ribs do not cool well and that these areas may require a longer cooling time.

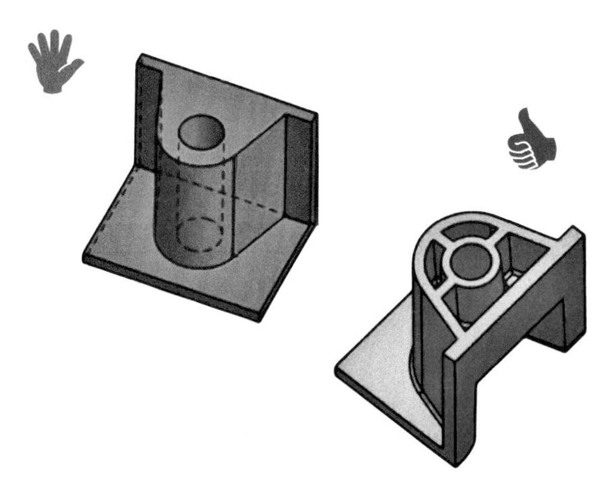

Figure 1.30 Connecting screw bosses to side walls [image source: Ritter/HS-Reutlingen]

1.2.1.5 Stiffeners

Warpage of a component means deformation after ejection

Plastic components can distort or warp after demolding (Figure 1.31). The reason is the plastic's low Young's modulus, i.e. the component is of low stiffness and possible process-related residual stresses can distort it. This is particularly problematic for flat components. Crowned components with slightly outwardly curved surfaces (crowning) are automatically stiffer and resist the residual stresses caused by the manufacturing process.

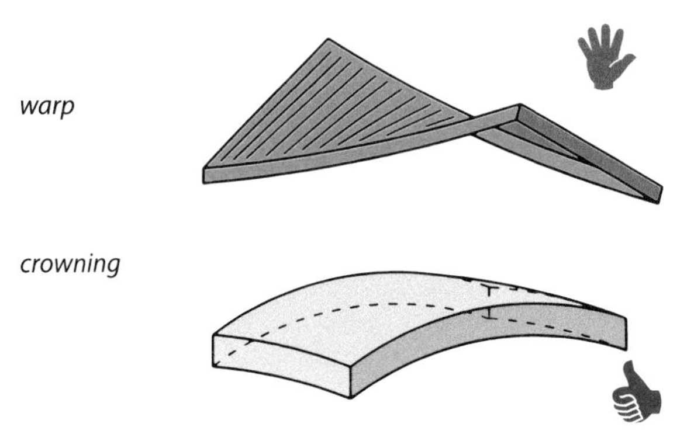

Figure 1.31 Planar components can warp, whereas crowned components have a “basic stiffness ”

Beads increase the stiffness of components

Beads are contour projections that also stiffen surfaces. In metal processing, for reasons of manufacturability, stiffening is usually done with longitudinal grooves or flutes, e. g. in a tin can made of tinplate, where the stiffening occurs in the direction of the beads. Plastic parts offer much more scope for designing beads. They can also be designed as patterns or lettering, in which case the bead is virtually imprinted, i. e. it can be seen on both sides of the component (Figure 1.32).

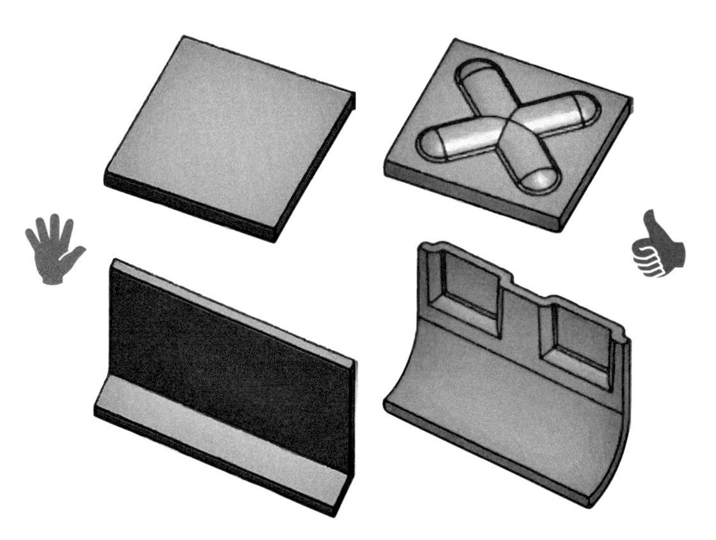

Figure 1.32 Stiffening of planar geometry by means of beads [image source: Ritter/HS-Reutlingen]

Finally, there is ribbing (Figure 1.33). Here, as is the case for beads, the stiffness is increased in the direction of the rib. At this point, it must be clarified:

- Is the rib merely to increase the stiffness of the component to counteract warpage? If so, the rib has to be quite thin so that no sink marks are formed on the opposite side.
- Is the rib intended to increase the strength of a component, e.g. for use in the base of a swivel chair? If so, the ribs have to be quite thick.

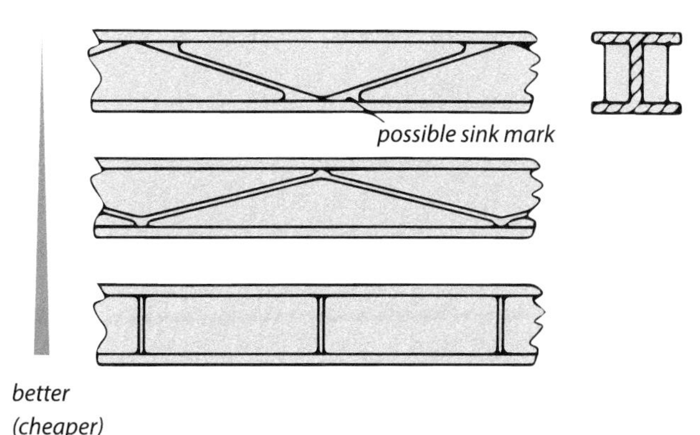

Figure 1.33 Design of ribbing

Thick ribbing stiffens a component, but can also be the cause of sink marks

To increase strength, the ribs must be thick and should be aligned with the force curve of the load situation. In this case, sink marks are to be expected on the opposite component surface, because material accumulation with a higher shrinkage potential occurs at the area in contact with the component surface. Countermeasures for this could consist in stiffening the surface by a strong convex curvature or a suitable surface structure to conceal the possible sink marks.

General design rules for ribs

If the ribs are merely intended to increase the inherent stiffness of the component, simple ribbing (Figure 1.33, bottom) is a more cost-effective mold design and more favorable for production. The diagonal ribbing shown results in very thin cores at the crossing points, which are unfavorable when it comes to cooling. Rib designs for component stiffening should take the following into account:

- Rib height at least 2.5 times the thickness of the area to be stiffened
- Demolding draft on both sides
- Rib thickness at the base should, as far as possible, be less than 75% of the thickness of the surface to be stiffened

1.2.1.6 Dimensional Change due to Temperature Fluctuations

The coefficient of linear expansion of plastics is about 10 times that of metals. This value indicates the change in length per change in °C. In the case of plastics, a component 1 m long will undergo a change in length of roughly 1 cm for a temperature change of 100 °C.

Many plastic components are part of assemblies of different materials. If these assemblies are intended for outdoor applications, the temperatures mentioned above may well be realistic. In winter, some countries can experience temperatures below −30 °C while, in summer, temperatures of 70 °C are possible behind a sunlit pane of glass.

The designer should therefore make allowance for the different linear expansions. For example, a component might be fixed at one end only and slide at the other. Here, the fixing points on one side could be designed as oblong holes.

1.3 Dimensional Deviations between CAD and Injection Molded Part

Injection molded parts differ fundamentally from their CAD geometry in terms of dimensions due to shrinkage of the plastic and possible warpage of the component.

Shrinkage and warpage lead to dimensional deviations

Shrinkage refers to the change in volume of a plastic during cooling. Warpage occurs when stresses in the component are greater than its inherent stiffness and lead to unwanted deformation.

Accurate prediction of the effect of shrinkage and warpage is almost impossible without simulation. Optimization loops between mold design and sampling in injection molding production are therefore common. They incur costs and can delay the start of production.

1.3.1 Shrinkage

Shrinkage occurs when the holding pressure is insufficient

If a component has different wall thicknesses, sink marks are very likely (Figure 1.34). The plastic will shrink during cooling and, where a large volume is involved, the absolute shrinkage volume will be greater. Because the cooling plastic becomes more and more viscous, the holding pressure cannot supply sufficient material, especially in areas remote from the injection point, and so material is missing here. Shrinkage generally occurs when the holding pressure is insufficient. For reasons related to the limited clamping force of an injection molding machine, this holds for almost all components. Only in the case of very small components (mini-injection molded parts) is the clamping force of the machine almost always more than sufficient, so that such small applications can be produced largely without shrinkage. There is even the possibility of mold overloading, whereby the plastic expands at the mold opening and the component is then larger than the cavity. Such overloading is also known to occur on normal-sized injection molded parts, but, if so, only locally near the injection point. If the ribs here have too small a draft, demolding difficulties can arise because a high holding pressure has overloaded the rib geometry.

Instead of sink marks, shrinkage can also lead to shrinkage cavities on the inside. This is always the case when the component surface is stiffened by an external bulge. Provided the components are not transparent, the shrinkage cavities will not be noticeable initially. However, they can lead to component failure under load.

Blowholes are quasi "internal sink marks"

If wall thickness differences cannot be avoided, the thicker area should always be close to the sprue.

Thick component areas should be positioned as close to the sprue as possible

Thick areas can be cored out many times, i.e. part of the volume is formed by a core in the mold so that the component becomes hollow in these areas.

Wall thicknesses become more uniform through coring

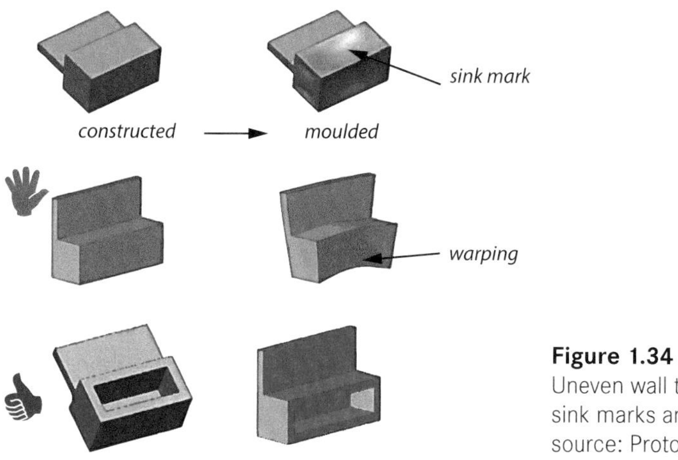

Figure 1.34
Uneven wall thicknesses may lead to sink marks and/or warpage [image source: Protolabs]

The component manufacturer can influence shrinkage, although there are natural, physical limits:

Shrinkage can be influenced via the process settings

- An overall high holding pressure leads to lower shrinkage. Usually, however, the available clamping force limits the holding pressure level.
- Particularly in the case of semi-crystalline plastics, shrinkage is greater with high mold temperatures because crystallization compacts the material and reduces the component's volume. However, it would be wrong to demand low

mold temperatures, because the plastic can still recrystallize in the application at high temperatures, with a subsequent change in component geometry.

Shrinkage must be factored into mold design

Shrinkage is predominantly a problem for the mold maker, because he/she has to enlarge the cavity for the component to account for the shrinkage. If the shrinkage value is not known exactly beforehand, the mold must be dimensionally designed in such a way that the required nominal dimension can be achieved iteratively by remachining after sampling.

The component will shrink onto mold cores during cooling, and, anyway, a high force will be required for the start of ejection, depending on the draft of the mold. Shrinkage is not merely an issue for isolated mold cores. Even where several cores are adjacent to each other, the plastic component will be unable to shrink freely between the cores (see Figure 1.35). The designer should keep this in mind and either plan for larger demolding angles or provide additional ejectors for this area.

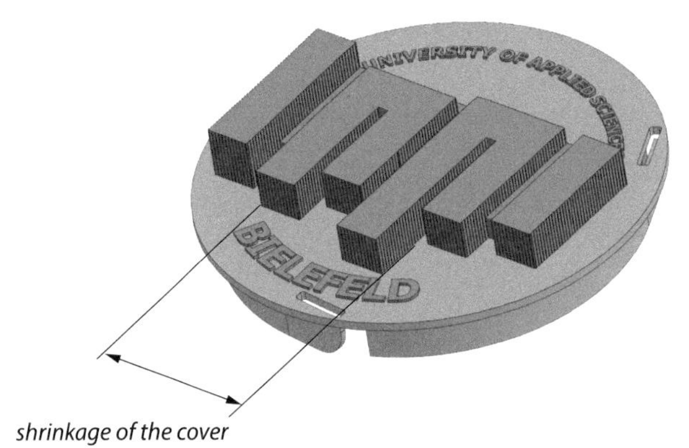

Figure 1.35
The area between protrusions cannot shrink freely
[image source: H. Peitzmeier]

Shrinkage values are not material constants – they also depend on the process control

The raw material suppliers provide shrinkage values for the plastics, but these values should be used with caution, because shrinkage depends not only on the material but also on the manufacturing process. The higher the holding pressure, the more plastic can still be pressed into the cavity during cooling, thus reducing shrinkage. This holding pressure effect is of course not isotropic, i. e. uniform in all spatial directions. Areas remote from the sprue do not feel the full effect of the holding pressure. The mold geometry can hold the plastic in place and thus hinder shrinkage. From this point of view, a local shrinkage value would have to be specified for each of the different mold areas, but this is hardly feasible in practice.

Shrinkage leads to two frequently occurring effects:

Sharp-edged component areas often have sink marks parallel with the edge

- *Edge collapse:* Shrinkage is hindered if the component shrinks onto a core and if parts of the melt solidify particularly quickly. The latter occurs at sharp-edged corners of somewhat thicker components. The still-hot melt inside the component continues to shrink during cooling, while the edge is already solid and resists shrinkage. In such cases, the edges can easily protrude (Figure 1.35).

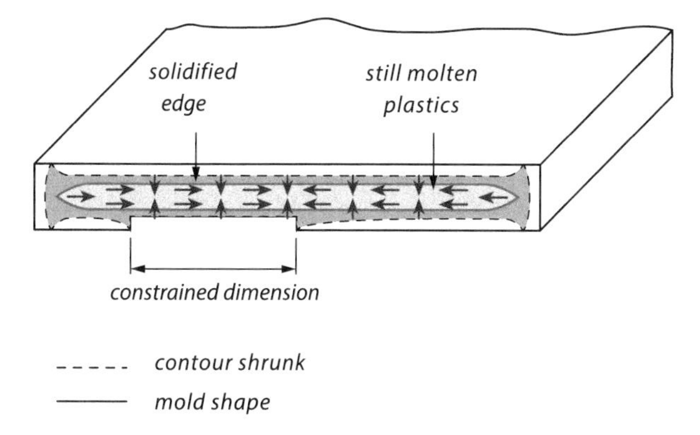

Figure 1.36 Shrinkage, free and constrained [image source: Stitz]

- *Corner collapse:* Box-shaped components often show inward deformation of the side faces. Although this is termed corner collapse, it refers to the warping of the adjacent surfaces. This phenomenon is related to insufficient cooling of the inner core. The outer surface has a large area and is surrounded by a large volume of tool steel, and so the core is often warmer (Figure 1.37). In the corner areas, a heat pocket forms closer to the core. Here, the plastic can shrink even further, generating a tensile stress that deforms the sidewall.

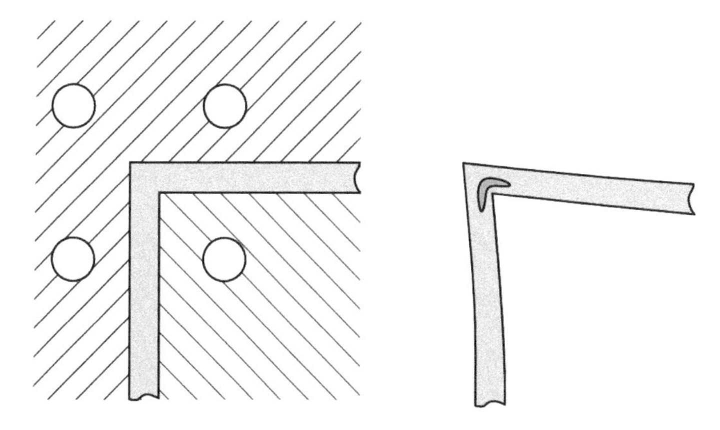

Figure 1.37 Corner distortion due to insufficient cooling of the internal corner

Distortion of adjoining surfaces often occurs at component corners

1.3.2 Warpage

Warpage occurs when the component stiffness is less than the internal stresses

Warpage always occurs when the internal stresses are greater than the component stiffness. Accordingly, a freshly molded component with its residual process heat has a high tendency to warp. With longer cooling times in the mold and thus under mold constraint, warpage is reduced, but this is not a reliable solution, because the stresses do not dissipate in a cold component and are thus conserved. The component will not until warp later, when it is exposed to heat.

Warpage must be factored into the design

Unlike shrinkage, the component manufacturer has only a few options for controlling warpage by changing the process control. In any event, it does not make sense to rely on special process control as a countermeasure for warped components, because this reduces the process window. The process window is understood as the bandwidth within which machine setting values may vary without affecting the quality of the components. The narrower the bandwidth, the higher the probability of bad parts that fail to possess the required quality. Now, if a simulation has shown that a specific mold temperature is required for preventing parts from warping, that exact temperature would have to be present in the mold during production, and that is very unlikely. First, the mold temperatures cannot simply be set and maintained in this way, because a constant temperature of the cavity in contact with the plastic is actually only established after a start-up process of approx. 20 cycles and thus depends on the cycle time, the amount of melt introduced and the heat transport capacity of the mold. Second, a different temperature may be necessary in production to achieve a defined surface gloss.

Causes of warpage

The warpage of a component is mainly predetermined by the design.

- Wall thickness differences cause different shrinkage.
- Ribs and beads increase component stiffness.
- Orientations, especially when using fibers, ensure very strong shrinkage anisotropy, i. e. shrinkage is lower in the direction of the fibers than across them.

Shrinkage not uniformly distributed causes internal stresses

Shrinkage can lead to warpage of the component after demolding. Shrinkage is specified as a percentage value. This results in different absolute shrinkage volumes for different wall thicknesses. In addition, shrinkage is not the same at all points of the component, because of local differences in holding pressure. This causes stresses in the component. In principle, thinner areas will be colder and therefore stiffer more quickly and will resist the shrinkage of the adjacent areas which are still thicker and warmer at the time of demolding.

Depending on the design, ribs can also promote warpage

If insufficient stiffness is provided by ribbing or crowning, the residual heat of the demolded component can lead to further contraction of the thick areas and thus to warpage (Figure 1.38). The warpage will mostly occur in the direction of the thick area, unless other component areas that are colder during demolding "counteract" it with even higher stiffness.

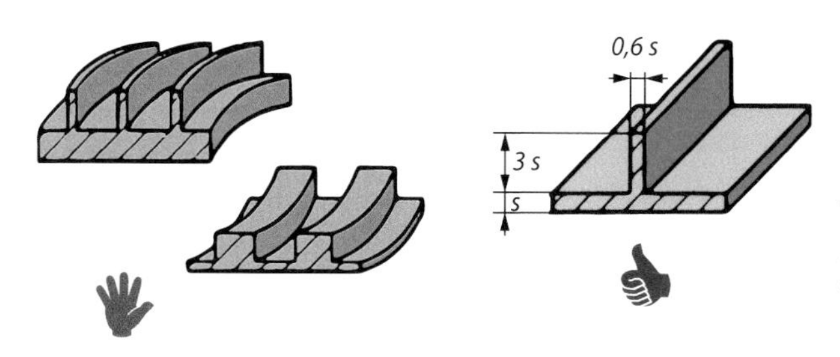

Figure 1.38
Warpage due to differential shrinkage at different wall thicknesses [image source: Meyer and Haak]

The flowing plastic is streamlined (see Chapter 2 "Manufacturing Process"). The alignment of the molecules leads to residual stresses during cooling, which can cause component distortion if the shape is not very stiff. This applies to flat and unribbed components. In many cases, however, the component design provides sufficient stiffness, with the result that residual stresses are not enough to distort the component.

Fiber orientation is a major cause of warpage

Very important, however, is the orientation of fibers that follow the flow. The fibers of filled plastics do not melt, of course, and so their change in length as a result of temperature changes during processing is also insignificant. However, the surrounding melt will shrink, as described earlier, and, in the direction of the fibers, shrinkage is significantly restrained. The outcome is substantial shrinkage anisotropy. Very often, the influence of fiber orientation on warpage is much stronger than all other design measures, i.e. ribbing, seepage and crowning. It is very important to consider component stiffness due to fiber orientations in the different spatial directions and to locally increase or weaken the component stiffness where simulation has revealed the probability of warpage.

1.3.3 Corrective Measures for Dimensional Deviations

Problem of predicting the influence of the process in the design phase

When designing an injection molded part, the influence of the process and its setting values should not be overestimated. Even with close coordination between designer, mold maker, and injection molder, it is not possible to clearly define the process parameters as early as the design phase, not least because no valid simulation models yet exist for surface effects such as gloss. From this point of view, all simulations should always be viewed with caution, because process settings have to be selected for running a simulation and these in the end will very probably not match the settings of the mold yet to be built.

Influence of the design on the dimensional accuracy of components

Favorable conditions for exact dimensions can be specified by the designer via:

- the choice of material and
- the determination of the shape

Fillers inhibit shrinkage and ensure better dimensional stability

The choice of material influences shrinkage. Fillers with a largely round shape reduce shrinkage considerably. This applies specifically to small glass beads added to the material as a filler. There are also materials that contain a mix of glass fibers and glass spheres, in which the fibers increase the stiffness the strength and the spheres reduce the anisotropy that occurs. Anisotropy means directional dependence – the fibers become aligned in the melt flow (Figure 1.39). Shrinkage is strongly reduced along the fibers, so that a round cavity inevitably produces more of an oval component.

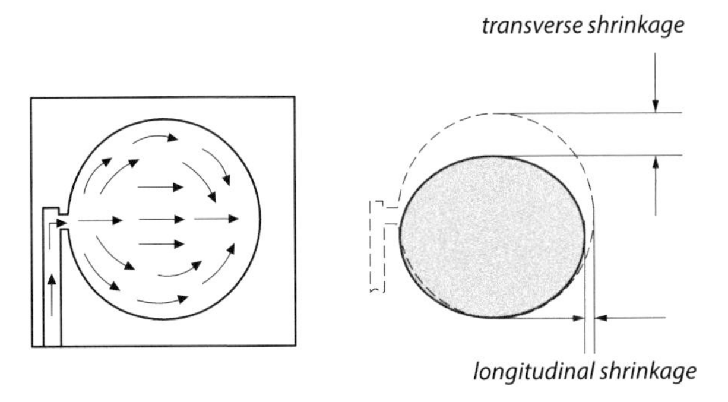

Figure 1.39 Anisotropy in fiber-reinforced plastics: longitudinal shrinkage is smaller than transverse shrinkage

At perpendicularly meeting melt flows and at flow end, fibers are predominantly oriented transverse to the flow direction (see Section 2.3, "Manufacturing Processes/Fiber Orientations"). The explanation for this lies in the flow behavior in a gap cross-section. In the area between the gate and the flow front, the melt flows laminarly at different velocities. In the center of the cross-section, the melt is hot and can flow faster here than at the boundary. At the flow front, the melt virtually swells between the solidified boundary areas into the cavity, while the longitudinally aligned fibers get automatically rotated. Consequently, this is a cause of warpage. Using the example of a flat sheet, it becomes clear that, at the end of the flow path, the fibers tend to lie parallel with the end of the sheet and thus hinder transverse shrinkage of the component. The component can now buckle or warp here (Figure 1.40).

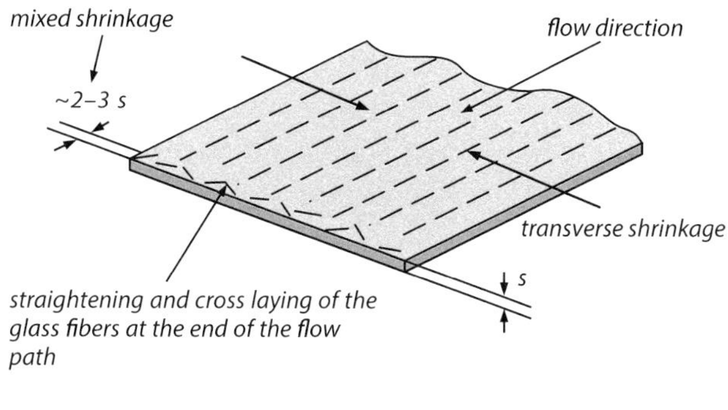

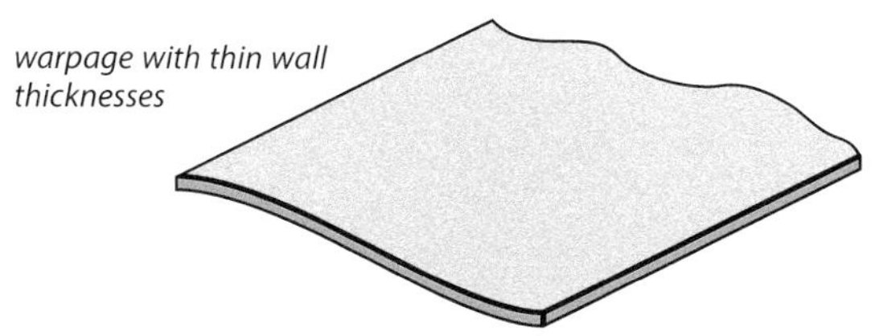

Figure 1.40
Warpage of a flat panel due to uneven alignment of reinforcing fibers [image source: Zöllner, Covestro/Leverkusen]

Design measures for counteracting tendency to warp

The designer has four options to reduce warpage:

- Increase component stiffness by ribs or beads
- Relocate the sprue and thus change the fiber orientation
- Influence the fiber orientation through targeted flow obstructions
- Anticipate warpage through negative correction

The sprue position determines the alignment of fibers

Figure 1.41 schematically shows the effect of a sprue gate. The inflow of the melt through a tunnel or sub gate gives rise to radially diverging melt flow. Shrinkage will be greater at the rear end of the component, causing it to warp. If the gate is a film gate, the melt can flow largely parallel through the cavity. The anisotropy is not eliminated, but the part is straight when it exits the mold.

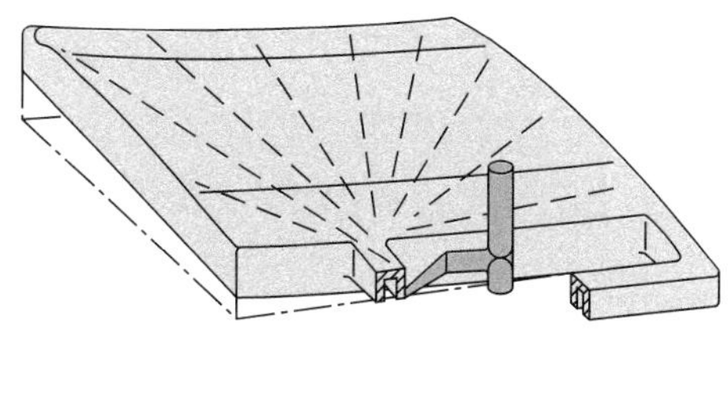

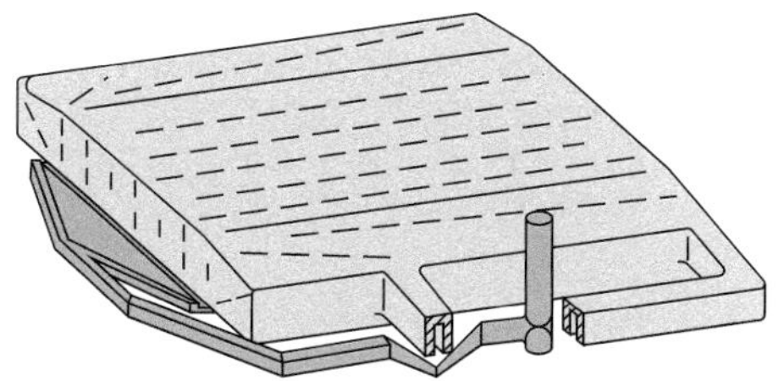

Figure 1.41
Warpage due to fiber alignment and correction by relocation of the sprue [image source: Zöllner, Covestro/Leverkusen]

Local influence of flow obstacles on fiber position and component stiffness

Flow obstacles can be used to change the fiber orientation. For example, this can be done by slightly tapering those wall thicknesses at which the melt is flowing more slowly overall, because the melt has a lower flow resistance in thicker areas and can flow faster (Figure 1.42). The flow obstacles therefore make it possible to change the fiber orientation and thus to locally reduce the stiffness in the flow direction.

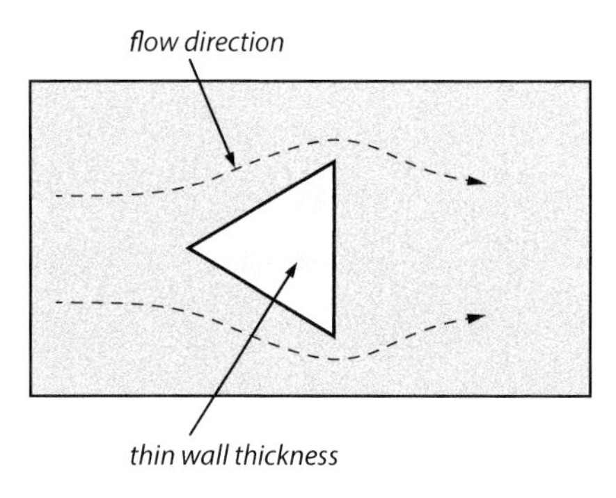

Figure 1.42
Change of fiber orientation due to flow obstacles

The effect of the above measures is hard to predict without simulations. The designer should proceed with great caution when using fiber-reinforced plastics and should simulate the possible warpage first. He/she should then use the results to make design changes and recheck them with new simulations.

Negative correction or dimensional hold-off

Designers with a very good knowledge and experience of simulations can also use the simulated warpage to effect a negative correction of the geometry. In Figure 1.43, the warpage is scaled up for clarity, the simulated deviations are multiplied by a factor and thus very exaggerated in the image. Based on this simulation, the component is bent in the opposite direction by the calculated value. In Figure 1.44, the lower right corner of the component has been shifted downward by 0.7 mm. This geometry forms the cavity of the mold. The process will subsequently cause the component to warp upward by the calculated 0.7 mm and then to have the originally intended shape.

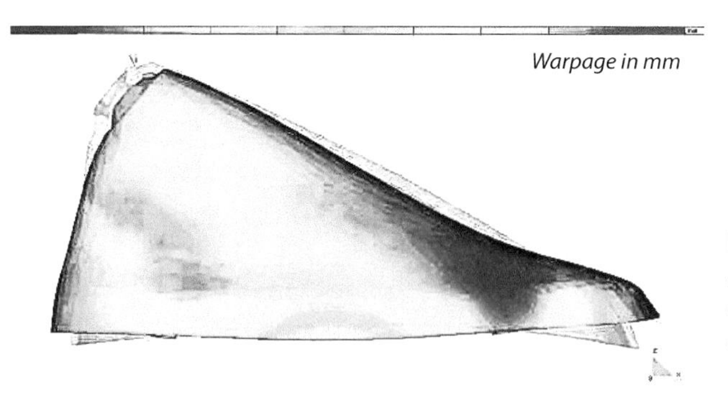

Figure 1.43
Simulation of warpage and comparison with CAD outline [image source: KB Hein GmbH]

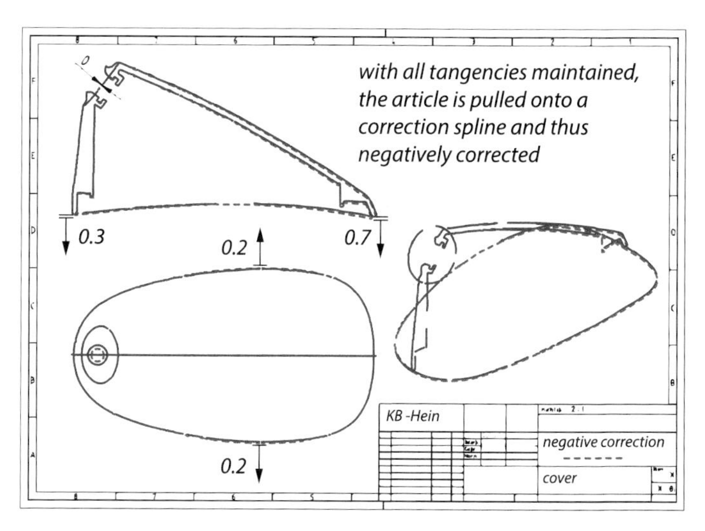

Figure 1.44 Negative correction of a component (dashed in red) [image source: KB Hein GmbH]

Negative correction can create undercuts

Negative correction of geometry based on simulated warpage is not without its problems. Simulation programs can automatically export a negatively corrected geometry, but they cannot know whether undercuts are created by the deformation of the component. The designer must therefore decide for him-/herself where to make the correction. If, for example, in Figure 1.44, the warpage calculated for the right component edge is 0.7 mm upward, he/she can move this edge of the component 0.7 mm downward and bind all neighboring areas with splines or tangents. If the subsequent warpage is 0.7 mm upward as calculated, the dimension will have been corrected exactly to 0.

In the ideal case, all the above measures result in a stiff contour that does not yield warped components even if the process settings are changed.

1.4 Design of Connections

The individual parts of an assembly can be connected by

- Screw fasteners
- Snap fits
- Bonding
- Welding

Each of these variants entails special design features. Because injection-molded parts are usually produced in large quantities, changes after the manufacturing of the injection mold are often difficult to implement and possibly also expensive.

1.4.1 Screw Fasteners

Screws for occasionally loosened connections

A screw fastener is always useful if the connection is to be loosened again. This is the case, for example, with housings that have to be opened occasionally for maintenance purposes. Basically, there are two possibilities:

- Use through-bolts and locknuts
- Insert a thread-cutting screw through one of the parts to be screwed together and screw it into a boss of the second component

In contrast to designs made of metal, the screw boss variant is very common for plastic components, because metal screws can cut a thread in plastics without any problems and because the screw fastener is thus only visible on one side of the assembly. This chapter will only cover screws made of metal and the general design of screw bosses (Figure 1.45).

Figure 1.45 Screw boss designs [source: Ehrenstein, G. W., *"Mit Kunststoffen konstruieren"*, 3rd ed., p. 153]

General design of screw bosses

Important for the design of screw bosses:

- In the case of lateral stabilization, the screw boss can be connected to side walls by means of thin webs.

- Despite the necessary draft angle, the base of the screw boss should not be too thick, as otherwise material will accumulate and sink marks can be expected on the opposite surface.

Depending on the torque, high axial forces can occur during as the screw is being tightened, which gives rise to a higher long-term static load on the plastic. Plastics can creep under such conditions, as a result of which the tightening force would decrease over time. It is possible to reduce this effect by increasing the contact area and thus reducing the stress in the plastic part. Figure 1.45 shows various designs for this. At the top left, a second boss has been constructed concentrically around the screw boss. The tightening force is transmitted to the two cylindrical bosses, thus reducing the stress when the screw is tightened very firmly. At the bottom left, the ribbing to the component side walls serves to absorb the tightening force via an additional pressure plate.

Plan screw bosses with low axial forces if possible

To ensure that the screw bosses can be made thin, special screws should be selected, where possible. These screws have very sharp flanks, with angles as small as possible below 30°, which make it easier to cut into the plastic. Furthermore, the acute angle reduces the radial force on the boss, as evidenced by the reduction in the force exerted by the flank on the plastic (Figure 1.46).

Screws for plastic joints have acute flank angles

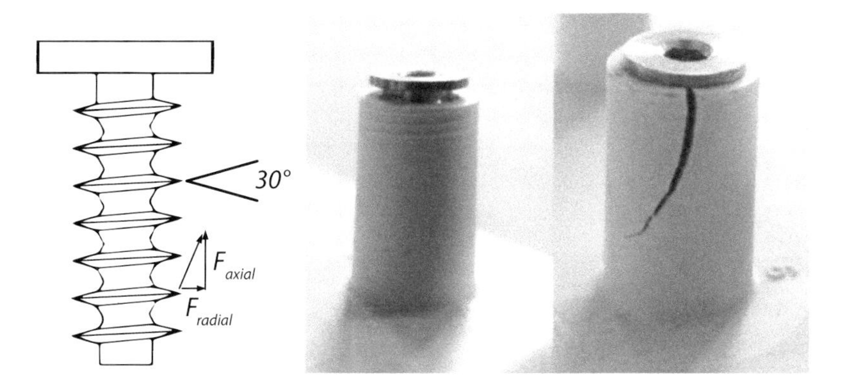

Figure 1.46 Screw and boss

The inner diameter and the wall thickness of the boss depend on the choice of screw. Here, the recommendations of the screw supplier should be followed. Regarding the inner diameter, the volume between the screw core and the boss must be large enough to be able to absorb the excess material produced when the screw is first screwed in. The inner diameter of the boss must also not be too large, so that the thread flanks can transmit sufficient axial force.

Follow screw suppliers' recommendations on screw boss geometry

Screws are best not pointed at their tips, because the screw bosses already have holes and, unlike wood or sheet metal screws, the screws do not have to create a

Screws with blunt tips

hole first. The blunt termination of the tip is also intended to ensure that the screw flanks, during a repeat tightening process, again find the thread that was cut during the first screwing process, if possible, and that they do not cut another thread. That would lead to destruction of the inner contour with each further turn of the screw and the screw would eventually no longer find a hold.

Screw inserts for frequently loosened connections

If a screw connection is to be loosened again and again, screw inserts can also be used. These are metal sleeves with, for example, metric internal threads. Different versions are available, which are either inserted into the mold and firmly bonded to the plastic during the subsequent injection process (Figure 1.47) or else are they are pressed in afterward, in which case, depending on the design, the plastic part can be fused on locally.

The provision of screw inserts automatically increases both outlay and costs. The overmolded variant produces a very tight connection. Handling robots are required for production and must be capable of gripping and positioning the inserts.

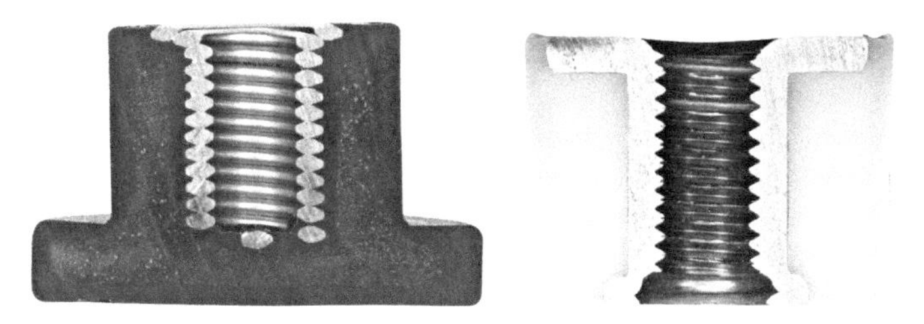

Figure 1.47 Screw inserts for metric screw fittings [image source: Böllhoff]

1.4.2 Snap-Fit Connections

Snap-fit connections require at least one elastically deformable mating partner

Snap-fit connections are very cost-effective because no additional elements are required. The elements required for functionality form part of the component design. This is a positive connection in which at least one of the mating partners has to be elastically deformed (Figure 1.48).

There are no constraints here at all on design, and the connections can be designed to be freely accessible for easy release. However, the design may also feature snap-fit hooks and latching grooves that are hidden inside a housing and are difficult to release.

Compared with a screw-fastener connection, the snap-fit connection is not as reliable, because plastics, with their low modulus of elasticity, are easily deformed and so the connection might loosen unintentionally when a load is applied.

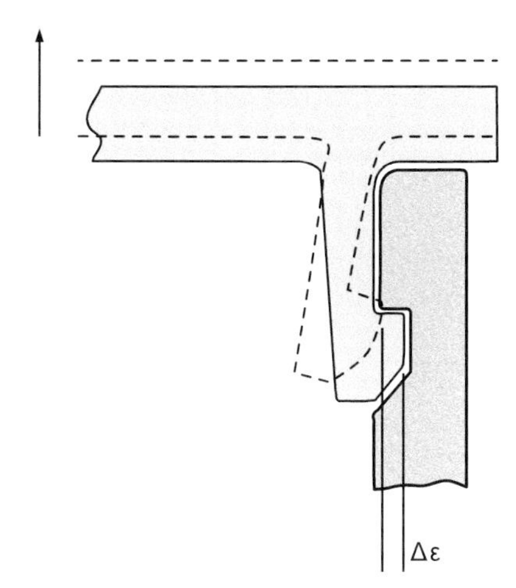

Figure 1.48
Snap-fit connection

The choice of material is important for the design here. Depending on the shape, a large deformation $\Delta\varepsilon$ occurs during mating. Amorphous plastics tend to be brittle and might break. Semi-crystalline plastics are viscoelastic above their glass transition temperature and can withstand large deformations in the short term. After mating, the deformation should be recovered in full and the mating partners should relax, because plastics can creep. In the event of creep, the residual stress will always disappear over time, but the snap-fit hook or the mating partner will be permanently deformed.

Simulations of snap-fit connections require "the right" characteristic values

An estimate of the maximum stresses using the calculation formulas employed in engineering mechanics [free calculation programs: Snap-Fit Design Calculator (BASF); FEMSnap Tool (Covestro)] shows whether the shape is possible with the selected material. The estimate of the resulting stresses is contingent on the correct choice of characteristic values. Because the mechanical behavior of plastics is strongly temperature-dependent, the critical failure state must be clear:

- If snap-fit connections are predominantly force-less, failure may occur during assembly. The following failure scenarios exist:
  - Failure during assembly
    Assembly in production takes place at room temperature (approx. 20 °C) and with rapid loading (rapid impact loading). For the calculation, the max. tolerable stress (yield stress or breaking stress) at room temperature is necessary. Because of the rapid loading, a material with the highest possible impact energy must be selected.
  - Failure in operation
    Actuation takes place under undefined conditions, which would be the case, for example, with a snap-fit belt buckle. The critical condition here would be the lowest temperature that might reasonably be expected on a winter's day. The stress that can be borne at −20 °C would have to be selected. At higher temperatures, the material will tend to be tougher and thus less likely to fail.

Failure in event of non-permissible deformation under load

- If the joint is subjected to temporary loads in the application, e.g. a snap-fit hook, it is more likely to fail at higher temperatures. Particularly in the case of semi-crystalline plastics, Young's modulus decreases significantly with temperature. In the case of plastics that can absorb moisture (especially polyamide), Young's modulus additionally decreases with increase in water content. Here, Young's modulus should be selected for the highest temperature that might realistically be expected and, if necessary, the highest moisture content.

1.4.3 Bonding and Welding of Seams

If a connection is to be inseparable, the components can be welded or bonded. Basically, the more dissimilar the materials are, the more problematic welding will be, because different plastics usually cannot be welded together. Compared with screw-fastened connections, adhesive-bonded and welded joints are more favorable in terms of mass production and the automation effort required.

1.4.3.1 Adhesive-Bonded Joints

Almost all plastics can be adhesive-bonded after pretreatment

Most plastics can be adhesive-bonded, but POM and polyolefins (PE and PP) are problematic. In many cases, pretreatment of the mating partners is necessary, i.e. surface changes induced by a plasma (ionized gas) or a locally limited thermal load (open flame, electric sparkover (corona treatment)).

Adhesive concepts

Whether pretreatment is necessary depends on the plastics themselves and the adhesives. The choice of adhesive itself also depends on the plastics; a rough distinction is made between:

- Solvent-based adhesives: The solvent should facilitate flowability of the adhesive so that it can be distributed as well as possible in the mating zone. The solvent can also slightly dissolve the plastic so that it swells somewhat, allowing adhesive to penetrate the top layer.
- Hotmelt adhesives that become flowable and develop their adhesive effect through the introduction of temperature.
- Chemically curing adhesives that can crosslink. These adhesives have good flowability before the chemical reaction. This group includes the two-component adhesives, whose two components react chemically, and the highly effective cyanoacrylate adhesives, which, however, have little moisture and temperature stability.

The variety of adhesives and their fields of application are very extensive and are not presented here in their entirety.

Load the bonding surface over as large an area as possible

In principle, an adhesive is weaker than the mating partners. Therefore, the areas to be bonded should be as large as possible. Because of the limited strength, an adhesive bond should always be loaded over its entire surface, if possible. Figure 1.49 illustrates the effect of a planar load compared with a linear load. From experience, a pressure-sensitive adhesive label is removed from a surface by generating a linear load, as shown in the lower part of Figure 1.49.

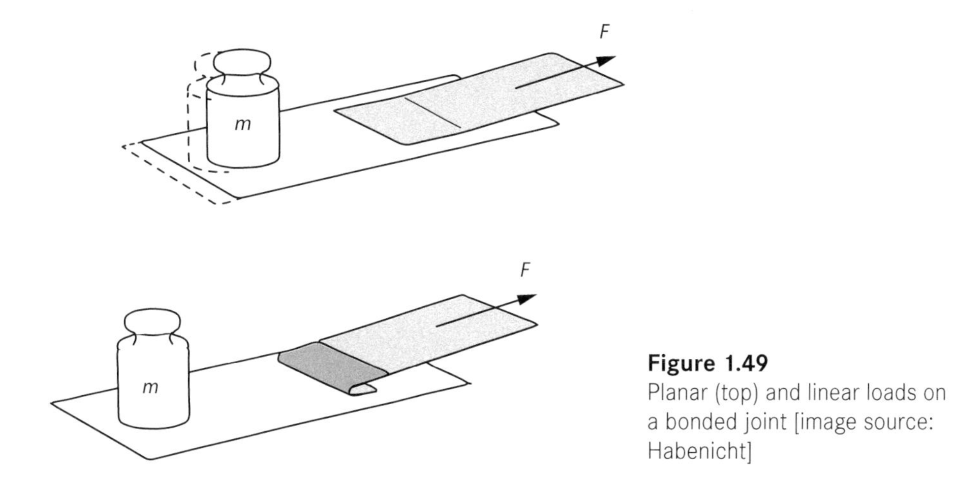

Figure 1.49
Planar (top) and linear loads on a bonded joint [image source: Habenicht]

When different materials are bonded, the mating partners may have different stiffnesses, either because the Young's moduli differ or because their shape differs. Under a shear load (Figure 1.50), the greatest component deformation and thus the greatest load on the adhesive will occur at the point of least stiffness, with the result that the bond may fail there.

Different values for Young's modulus may lead to load peaks

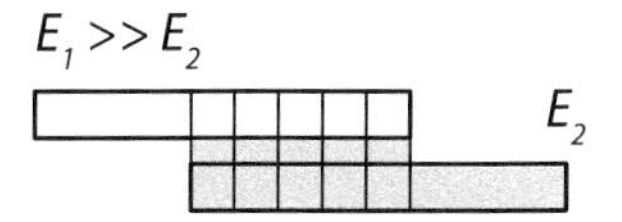

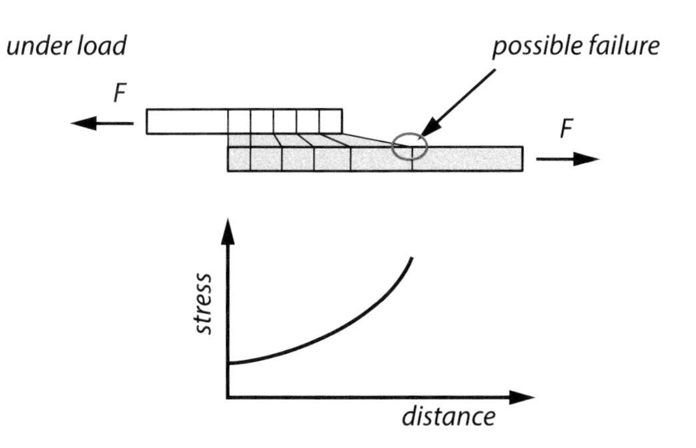

Figure 1.50 Stress distribution in an adhesive-bonded joint

Design features of adhesive-bonded surfaces

The design of an adhesive-bonded joint should factor in the following:

- The mating zone should be as large as possible.
- The mating zone should be loaded as uniformly as possible; linear loads should be avoided.
- The component stiffness may have to be increased in the area of the joint, e.g. by ribbing or by thickening over several layers bonded on top of each other.

1.4.3.2 Welded Joints

Two plastic melts can flow toward each other and the molecules can hook into each other like a hook-and-loop fastener. Different plastics are usually immiscible with each other, which is why they cannot be welded together. Consequently, the choice of materials for a welded joint is very limited. Table 1.3 shows a rough overview.

Table 1.3 Overview of Possible Welding Partners

	PE	PP	ABS	PS	PBT	PET	PA	PA6.6	PC	PC/ABS	PMMA	POM	SAN	TPE-U	TPE-S
PE	3						1								
PP		3					1					1			
ABS			3		3				3	3	3		3	3	
PS				3											2
PBT					3				2				3		
PET						3									
PA							3	3						2	
PA6.6								3							
PC									3		2				

Weldability:
3 = Good
2 = Poor
1 = For special compounds containing appropriate additives for increasing compatibility

There are various welding processes available for plastics, each involving a different type of fusion. Each requires special design component features. For injection-molded parts, the most common processes are:

- Hot plate welding
- Infrared welding
- Friction welding
- Ultrasonic welding
- Vibration welding
- Laser beam welding

In hot plate welding, the components to be joined are brought into contact with a hot plate until the mating surfaces melt, after which they are pressed together. This requires machines that can press the mating surfaces against the hot plate, move the hot plate to one side and then press the components together. The mating surfaces of the components must be mirror-symmetrical to each other and level (Figure 1.51).

Hot plate welding or hot plate butt welding

Infrared welding is very similar to hot plate welding, but the heat source is an infrared radiator. Curved glass tubes with internal glow wires are used. These emitters can be manufactured specifically for an application and do not necessarily have to be flat. Therefore, the mating surfaces only have to be mirror symmetrical with each other.

Infrared welding

The two processes, hot plate and infrared welding, each require certain travel distances of the components because the hot plate or infrared emitter must be moved out of the way again before the actual mating process. During the time of travel, temperature equalization occurs, and to ensure that the temperature is still high enough for welding, a relatively large amount of plastic has to be melted. The other processes manage without these transfer movements, so that less plastic needs to be heated.

Friction welding uses as its heat source mechanical energy which is generated when two components rub against each other. In the simplest case, these are two rotationally symmetrical bodies. A mating surface in the form of a V-groove is very suitable here, since in this way the components center independently. One of the parts to be mated rotates about its axis of symmetry, while the counterpart is pressed against it under slight pressure (Figure 1.51).

Friction welding

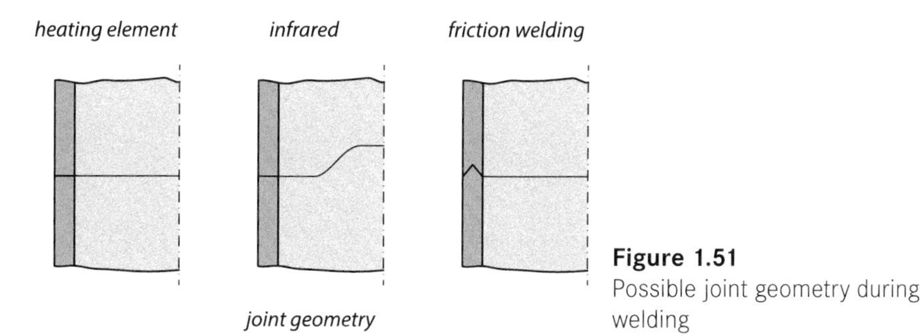

Figure 1.51
Possible joint geometry during welding

In ultrasonic welding, a high-frequency vibration is applied to the component by means of a sonotrode. The vibration is small and low in energy and is transmitted via the plastic to energy direction transmitters (EDT) and virtually bundled. These EDTs are:

Ultrasonic welding

- Small cone-shaped elevations (Figure 1.52)
- V-shaped elevations

Regarding material selection, it is important to note that very soft materials are difficult to weld using ultrasound because the vibration is absorbed too much by the plastic and only a small part of the vibration reaches the EDT.

Vibration welding

Vibration welding is very similar to friction welding. Here, machines are used that generate low vibration in two axes. Again, one of the mating parts is stationary, while the counterpart rubs against the surface with a certain pressure and generates heat. The mating zone here can be a circumferential groove whose width is greater than the vibration path.

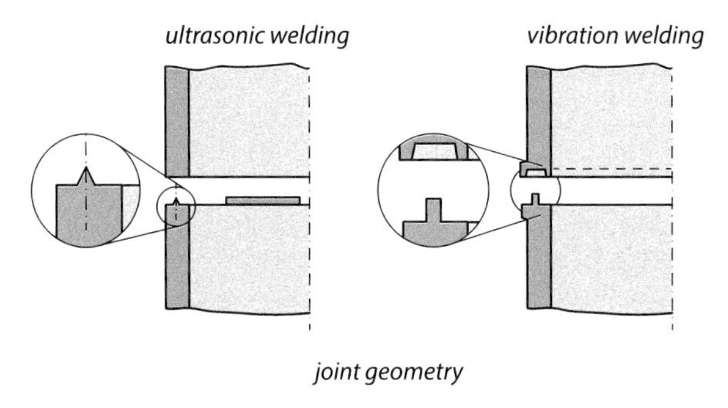

Figure 1.52 Joint geometry for ultrasonic and vibration welding

Laser beam welding

For laser beam welding, the energy for melting is generated by a laser. The components to be welded are first positioned relative to each other, then a laser beam passes through one of the components and melts the second component at the contact point (Figure 1.53). For this to happen, one component must allow the laser beam to pass through it, and the second component must absorb the laser beam completely. This is achieved with fillers added to the second material, i.e. the fillers essentially absorb the energy of the laser beam.

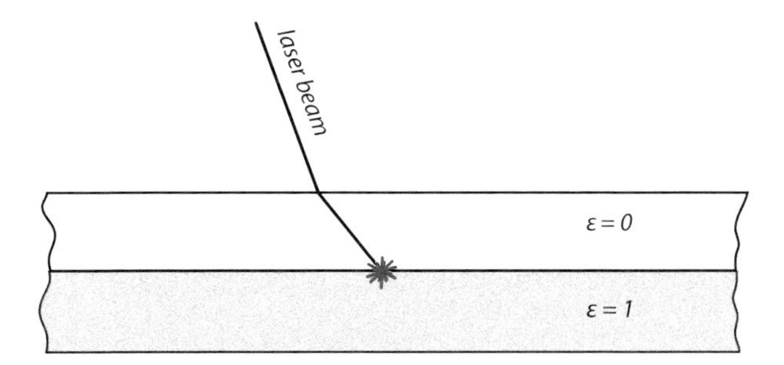

Figure 1.53 Principle behind laser beam welding

The mating surface can therefore be designed for laser beam welding without any special measures. At most, it must be remembered that the components must make full-face contact with each other and that the laser beam has to reach the contact point in the case of curved surfaces. There must therefore be no shadow areas for the laser beam, a constraint which means that deep, fissured geometries tend to be unfavorable.

1.4.3.3 Film Hinges

Film hinges connect two components during the injection process

Film hinges are movable connections between two components and are created during the injection process. In principle, a marked reduction in wall thickness constitutes such a hinge. In the design of a film hinge, the gate position is particularly important. If it is close to the hinge, when the melt reaches the constriction, it is slowed down to such an extent that it cools down too much because of the very small wall thickness in the hinge area. When the melt has filled the entire first area, the part of the melt that reached the hinge area last is still hot enough and will easily overcome the constriction. Thus, there is the risk of a weld line, because the melt flows into the thin area on both sides and comes together in the middle (Figure 1.54).

Gating point for the production of film hinges

The gating point should therefore be as far away as possible from the film hinge or, if possible, the melt flow front should reach the bottleneck at all points at the same time.

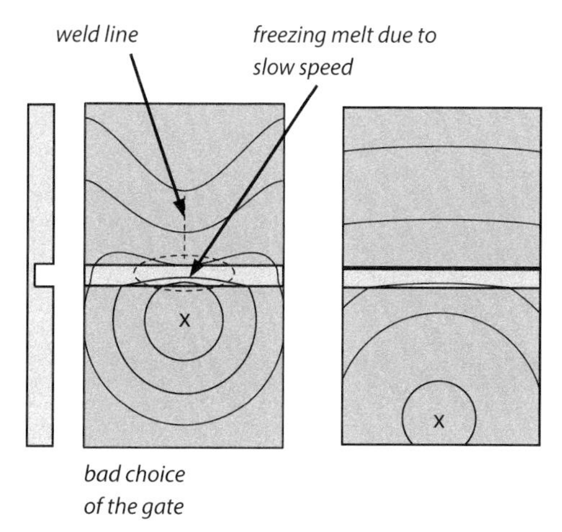

Figure 1.54
Filling pattern for a flow barrier (film hinge with different gates, gate positions)

Prerequisite for the functioning of film hinges

To ensure the mobility of a film hinge, a semi-crystalline plastic with a very low glass transition temperature must be selected, or the glass transition temperature must be below the application temperature. Amorphous plastics are therefore unsuitable for film hinges; only semi-crystalline plastics can be selected.

For a film hinge to withstand a great many movement cycles, the design is less essential. It is much more important that the plastic around the hinge be overloaded once beyond the mating limit (Figure 1.55). When the yield stress is exceeded, necking occurs in the tensile test followed by neck formation in the necking area (shoulder-neck necking). In the process, the molecules flow past each other and are able to align themselves somewhat.

Overstretching the film hinge once strengthens the plastic at that point

The technical stress in a tensile test is usually calculated by relating the force F to the initial cross-section A_0, because it is very difficult to measure the actual cross-section during the test. If, after exceeding the elastic limit, a plastic exhibits increasing necking with largely constant width, this is referred to as shoulder-neck deformation. In the area of the shoulder-neck necking, the cross-section barely changes at all, so it is quite easy to calculate the true stress here by relating the load F to the smaller cross-section A_{min}. The plastic can withstand considerably higher stresses after the yield stress is exceeded.

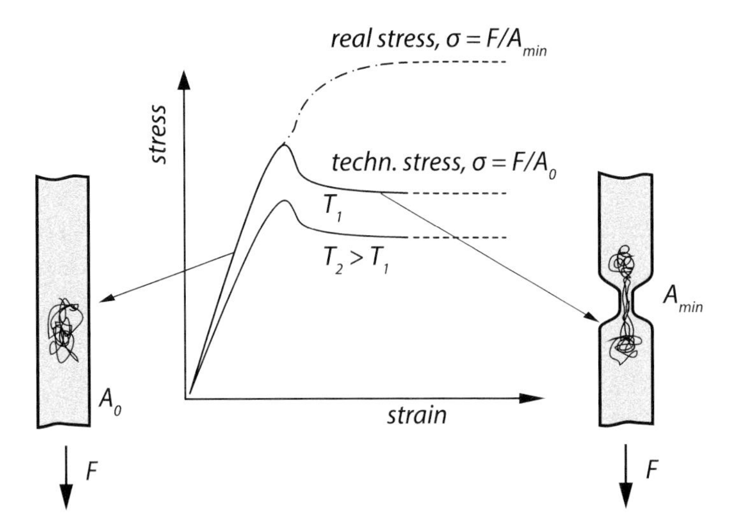

Figure 1.55 Alignment of molecules and increase in tolerable stress after yield stress is exceeded

Ideally, the first overload takes place at elevated temperatures, because then the yield stress for this effect is lower. Film hinges are often actuated for the first time before demolding while there is still residual heat in the mold; for example, in the case of closure caps, a flap mechanism integrated into the mold rotates the lids of the closure caps by 180° and thus closes the cap even before demolding.

1.5 Tolerances and Dimensions

Unlike metal components, greater deviations in dimensions, shape, and position must be expected with plastics. Consideration must also be given to the properties of the plastics after production, i. e. in the application (Figure 1.56). This means:

Plastic components have larger tolerances than metal components

- Immediately after demolding, the components are still hot, or at least residually warm. Reliable measurement is only useful once the components have cooled down completely, because the thermal expansion of plastic is large, and so a noticeable dimensional reduction occurs after demolding. The dimensional change occurring after demolding until cooling is complete is referred to as processing shrinkage.

Component dimensions during manufacture

- Post-shrinkage is a further dimensional reduction resulting from post-treatment; this essentially refers to heat treatment. Semi-crystalline plastics can still post-crystallize after a long time if the components are exposed to high temperatures.
- Some plastics can absorb moisture and thus become slightly larger. The moisture absorption varies with the ambient conditions, i. e. at low humidity the plastics release the moisture again and become slightly smaller.

Environmental influences on component dimensions during use

- Because of the strong thermal expansion of plastics, their dimensions change with change in service temperature. They can become even smaller at very low temperatures or larger at high temperatures.

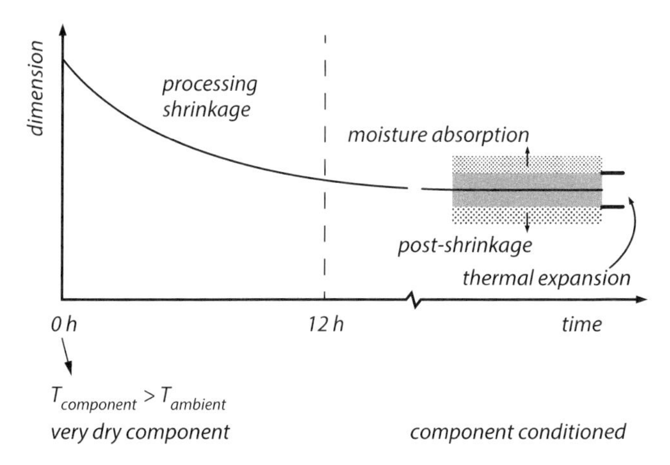

Figure 1.56 Dimensions after production

- The tolerance standards that apply to metals can therefore not be adopted for plastic designs, or, if at all, only to a very limited extent. The standard DIN 16742:2013-10 considers the interdependencies between component development, tool and mold making, and component manufacture.

DIN 16742 lists reasonable tolerances that vary according to the quality of the component manufacturer and the mold maker

- The molded part designer decides on the functionally required tolerances resulting from the application conditions and assembly.
- When the order is placed, the molding material specified by the molded part designer is binding. He/she thus lays the foundation for determining the processing shrinkage. After the order has been placed, the values for processing shrinkage have to be agreed between the molded part manufacturer and the mold shop or mold designer, possibly drawing on the experience of a third-party (e. g. the raw material producer).
- The molded part manufacturer confirms the allowable manufacturing tolerances for the acceptance conditions of the molded part production to comply with the relation "possible manufacturing tolerance ≤ functional tolerance", whereby financial agreements (e. g. price surcharges) may have to be included. The functionally required tolerances must always be specified in the design documentation. This avoids nonsensically precise and uneconomical tolerances based on fear or habit.

Component dimensions of plastic parts are also influenced by the quality of manufacturing

Depending on the material specification, component design and mold design, the processing of the plastics exerts a considerable influence on the dimensional accuracy of the molded parts. The processing machines employed in the primary molding processes are complex thermodynamic-rheological integrated systems which, despite highly developed manufacturing technology, are still largely operated and optimized empirically.

Dimensional properties of plastics include the extreme range of type-dependent stiffness or hardness and processing shrinkage. Transient and inhomogeneous mold and molding temperatures in conjunction with flow-related orientations of microstructures and additives lead to property anisotropies that cause pronounced warpage (warping, twisting, warping) of the molded parts. Furthermore, wall thickness differences or material accumulations can be possible causes of warpage. This is associated with form, position, and angle deviations in a highly complex manner which, compared to metals, make standardization considerably more difficult.

DIN 16742:2013-10 Plastic molded parts – Tolerences and acceptance conditions, Beuth-Verlag Berlin

Tolerances for proper functioning in assemblies are not specified in the standard

The tolerances that ensure the functionality of the components have to be defined by the molded part developer/designer. Standardization is not possible here due to the large number of applications! What the standard offers is the possibility of checking whether compliance with the tolerance of the manufacturing dimensions is technologically possible.

Furthermore, the standard can be used to estimate the necessary manufacturing effort. As in all other quality areas, the standard cannot protect against over-confidence.

In the case of components, a distinction is made between mold-specific dimensions and non-mold-specific dimensions (Figure 1.57 and Figure 1.58). By mold-specific dimensions is meant all component areas formed by more than one mold element. A slider that enables the demolding of an undercut is inevitably a moving element and thus the position of the axis of movement is no longer very precisely defined.

Component dimensions are mold-specific or non-mold-specific

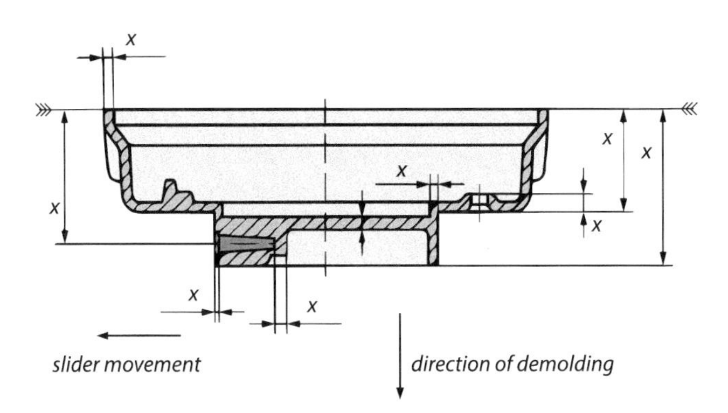

Figure 1.57 Non-mold-specific dimensions (DIN 16742)

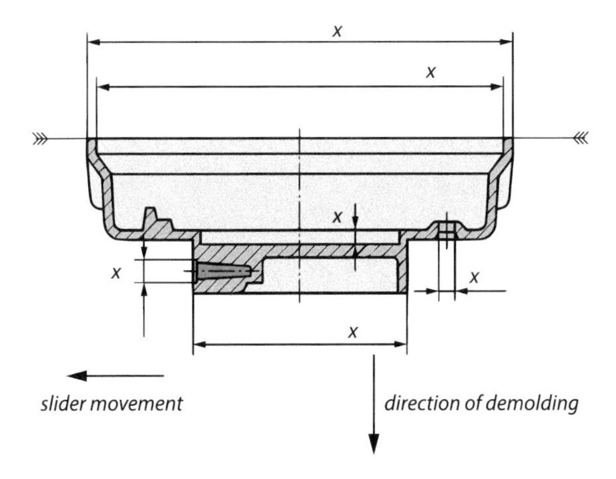

Figure 1.58 Mold-specific dimensions (DIN 16742)

For general tolerances, a distinction is made not only between specific and non-specific dimensions. In the current DIN 16742, the tolerances are listed in tolerance groups for different component dimensions, as shown in the table (Figure 1.59). The tolerance groups (TG) are not freely selectable, but instead result from the influencing factors described below. In contrast to metals, very small tolerances can be achieved for small plastic components with relatively little effort. The smaller the component becomes, the greater the influence of the precision of the manu-

General tolerances depend on the grouping into tolerance groups

facturing tool. The larger the plastic components become, the more difficult it is to maintain small tolerances.

Plastic moulded part tolerances as symmetrical limit dimensions for sizes (excerpt from DIN 16742)t

Tolerance group		Limit dimensions (GA) in mm for nominal size ranges in mm												
		1 to 3	>3 to 6	>6 to 10	>10 to 18	>18 to 30	>30 to 50	>50 to 80	>80 to 120	>120 to 180	>180 to 250	>250 to 315	>315 to 400	>400 to 500
TG1	W	± 0.007	± 0.012	± 0.018	± 0.022	± 0.026	± 0.031	± 0.037	± 0.044	–	–	–	–	–
	NW	± 0.012	± 0.018	± 0.022	± 0.026	± 0.031	± 0.037	± 0.044	± 0.050	–	–	–	–	–
TG2	W	± 0.013	± 0.020	± 0.029	± 0.035	± 0.042	± 0.050	± 0.060	± 0.090	± 0.13	± 0.15	± 0.16	± 0.18	± 0.20
	NW	± 0.020	± 0.029	± 0.035	± 0.042	± 0.050	± 0.060	± 0.090	± 0.13	± 0.15	± 0.16	± 0.18	± 0.20	± 0.22
TG3	W	± 0.020	± 0.031	± 0.05	± 0.06	± 0.07	± 0.08	± 0.10	± 0.15	± 0.20	± 0.23	± 0.26	± 0.29	± 0.40
	NW	± 0.031	± 0.050	± 0.06	± 0.07	± 0.08	± 0.10	± 0.15	± 0.20	± 0.23	± 0.26	± 0.29	± 0.40	± 0.55
TG4	W	± 0.03	± 0.05	± 0.08	± 0.09	± 0.11	± 0.13	± 0.15	± 0.23	± 0.32	± 0.35	± 0.41	± 0.45	± 0.63
	NW	± 0.05	± 0.08	± 0.09	± 0.11	± 0.13	± 0.15	± 0.23	± 0.32	± 0.35	± 0.41	± 0.45	± 0.63	± 0.88
TG5	W	± 0.05	± 0.08	± 0.11	± 0.14	± 0.17	± 0.20	± 0.23	± 0.36	± 0.50	± 0.58	± 0.65	± 0.70	± 1.00
	NW	± 0.08	± 0.11	± 0.14	± 0.17	± 0.20	± 0.23	± 0.36	± 0.50	± 0.58	± 0.65	± 0.70	± 1.00	± 1.40
TG6	W	± 0.07	± 0.12	± 0.18	± 0.22	± 0.26	± 0.31	± 0.37	± 0.57	± 0.80	± 0.93	± 1.05	± 1.15	± 1.60
	NW	± 0.12	± 0.18	± 0.22	± 0.26	± 0.31	± 0.37	± 0.57	± 0.80	± 0.93	± 1.05	± 1.15	± 1.60	± 2.20
TG7	W	± 0.13	± 0.20	± 0.29	± 0.35	± 0.42	± 0.50	± 0.60	± 0.90	± 1.25	± 1.45	± 1.60	± 1.80	± 2.60
	NW	± 0.20	± 0.29	± 0.35	± 0.42	± 0.50	± 0.60	± 0.90	± 1.25	± 1.45	± 1.60	± 1.80	± 2.60	± 3.50
TG8	W	± 0.20	± 0.31	± 0.45	± 0.55	± 0.65	± 0.80	± 0.95	± 1.40	± 2.00	± 2.30	± 2.60	± 2.85	± 4.00
	NW	± 0.31	± 0.45	± 0.55	± 0.65	± 0.80	± 0.95	± 1.40	± 2.00	± 2.30	± 2.60	± 2.85	± 4.00	± 5.50
TG9		± 0.30	± 0.49	± 0.75	± 0.90	± 1.05	± 1.25	± 1.50	± 2.25	± 3.15	± 3.60	± 4.05	± 4.45	± 6.20

W: tool-specific dimensions
NW: non-tool-specific dimensions
The differentiation of tool-specific and non-tool-specific dimensions is not necessary for TG9.

Figure 1.59 Tolerances for plastic components according to DIN 16742

Calculation recommendation for sorting into tolerance groups

The tolerance groups result from Figure 1.60 and take into account the manufacturing process, the materials processed and the production of the components themselves.

TG	TG1	TG2	TG3	TG4	TG5	TG6	TG7	TG8	TG9
P_g	1	2	3	4	5	6	7	8	≥ 9
total number of points $P_g = P_1 + P_2 + P_3 + P_4 + P_5$									

Figure 1.60 Determination of tolerance groups according to DIN 16742

The respective influences each have a points total, which is calculated as follows:

- P1 refers to the manufacturing process. The molding processes mentioned here relate to thermoset materials, which are not covered in this book.

Manufacturing process	P_1
Injection molding, injection compression, injection compression molding	*1*
compression molding, flow molding	*2*

- P2 considers the stiffness or hardness of the material. Stiff and very hard materials allow tighter tolerances and have a lower points value.

Material stiffness or hardness			P_2
Young's modulus in N/mm²	*Shore D*	*Shore A; IRHD*	
> 1200	*> 75*	–	*1*
> 30 ... 1200	*> 35 ... 75*	–	*2*
3 ... 30	–	*50 ... 90*	*3*
< 3	–	*< 50*	*4*

- P3 considers the calculated processing shrinkage, which should be agreed with the molded part manufacturer, if possible. This is where experience and the information provided by the raw material manufacturers count. The processing shrinkage is not a material constant; it also depends very much on the processing method, i.e. also on the level of holding pressure.

Processing shrinkage (calculated value)	P_3
< 0.5%	*0*
0.5% ... 1%	*1*
> 1% ... 2%	*2*
> 2 %	*3*

- P4 considers the range of variation in processing shrinkage due to molded part design and processing influences. This determination is not simple. The standard suggests setting the value to 1 if calculated values of the processing shrinkage are known, e.g. from experience, systematic measurements, or computer simulations, and if shrinkage anisotropy is insignificant or can be considered with sufficient accuracy in the respective dimensional direction, which unfortunately is rarely possible in practice. Where no exact information can be provided, the value 3 should be selected.

Consideration of geometry- and process-related shrinkage fluctuations of the VS calculated value	P_4
exactly possible: deviations from calculated value max. ±10%	*1*
possible with limited accuracy: deviations from calculated value max. ±20%	*2*
only possible inaccurately: deviations from the calculated value significantly above ±20%	*3*

- P5 concerns the production itself and must be coordinated with the component manufacturer. Smaller tolerances are possible if the effort for monitoring the machine function is increased in production and it is also ensured that changes in the actual process values have at least a monitoring of the quality as a consequence. Then a smaller tolerance group can be selected.

Tolerance series	P_5
Series 1 (standard production): Production with general tolerances may be possible. Dimensional accuracy requirements are not a particular quality focus.	*0*
Series 2 (exact production): Production and quality assurance are oriented towards higher dimensional accuracy requirements.	*−1*
Series 3 (precision manufacturing): Complete alignment of manufacturing and quality to the very high dimensional accuracy requirements.	*−2*
Series 4 (precision special production): like series 3, but with intensive process monitoring.	*−3*

The tolerance series 3 and 4 are always in accordance with the agreement.

1.6 Sizing

Dimensioning the wall thickness

For plastic parts, the determination of dimensions largely concerns wall thicknesses; the length and width are usually determined by the specifications or the requirements of the application. To ensure that the component can withstand the expected loads, the calculations typically employed in mechanical engineering are used.

Design limits are the stresses that can be borne

The design limits are the maximum stresses that can be borne. Whereas, in metal designs, the limit stress is the yield strength $R_{p0.2}$, or the yield strength R_e is selected, the choice for plastics is less clear (see Chapter 5 "Material Selection").

Choice of load limits depends on temperature at which damage is expected

- First, the temperature at which the component is to be used or will fail has to be clarified. The mechanical behavior of plastics is significantly influenced above the glass transition temperature as the temperature rises. Young's modulus becomes smaller at higher service temperatures, which means that the component stiffness is reduced. Fracture becomes increasingly tougher at higher temperatures under both static and dynamic loads (impact loads), and the maximum stresses achieved in the tensile test become smaller. If possible, the calculations should be carried out on characteristic values measured at temperatures at which possible damage is expected.

Load limits for short-time loading

- For short-term static loads, select the following limit stresses on the basis of how the material behaved in the short-term tensile test (Figure 1.60):

- **Amorphous plastics** break without prior notice or necking when the breaking stress is reached.
- **Semi-crystalline plastics** exhibit yield stress and subsequent necking in the range of the glass transition temperature and above. At progressively higher temperatures, plug-in stress decreases and the necking can degenerate to shoulder-neck deformation, with very large permanent strains becoming possible.
- Very **soft materials** have a very low modulus of elasticity, and so deformation becomes quite large even under low stress. In these applications, it is useful to set the stress limit at a value where the permanent elongation is 0.5%. Of course, larger values could also be selected, but then the component would deform by more than the desired amount under the load.

Load limits for static continuous load require isochronous stress-strain diagrams

- For long-term static loads, there are two possible damage cases; the design limits are selected in each case on the basis of isochronous stress-strain diagrams, which are based on the results of creep tests. Isochronous diagrams show the time-dependent strain under constant load. For explanations of these special diagrams, see Section 5.2.4.
  - Unlike metals, plastics can creep noticeably under continuous load even at room temperature, i. e. they can become longer. This creep is especially significant above the glass transition temperature, i. e. it particularly affects semi-crystalline plastics. For these applications, if possible, isochronous stress-strain diagrams should be used to select the limit stresses at which a permissible selected strain of ε_y after a selected load time t_x is exceeded.

    At this point, it must unfortunately be pointed out that these isochronous stress diagrams are available for very few plastics. In the case of a loading time of approx. 10,000 h, the limit stress from a usual short-time tensile test is selected and multiplied by 0.5 (Figure 1.61).

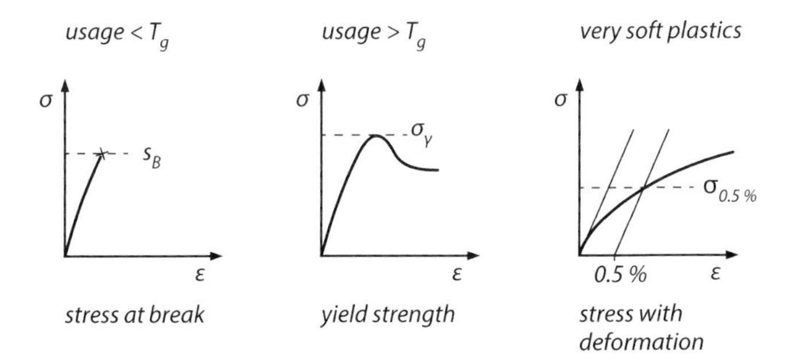

Figure 1.61 Limit stresses for dimensioning for short-term loads

Under continuous static load, amorphous plastics can develop visible but small cracks (crazes)

- Amorphous plastics, especially transparent ones, exhibit small cracks (crazes, see Section 5.3.2) when exposed to static loads for a long time. It has been proven that the mechanical properties do not suffer from this, but especially for transparent applications, e. g. a "viewing window", this would be an optically unacceptable defect. Investigations by Taprogge (1974) showed that these cracks do not form if the load is so small that the permanent deformation is less than 0.8%. A strain of 0.5% is assumed to be a safe limit value here, and is again taken from an isochronous stress-strain diagram (Figure 1.62).

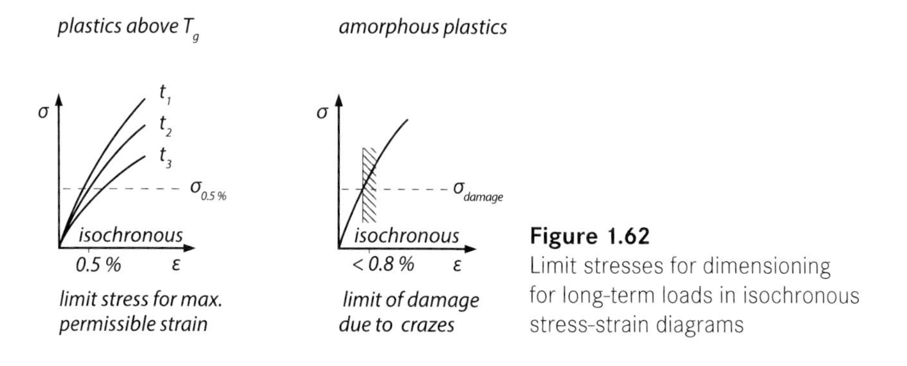

Figure 1.62
Limit stresses for dimensioning for long-term loads in isochronous stress-strain diagrams

Load limits for cyclically loaded components are not clearly defined.

- For cyclical loads, the procedure is not clear. Wöhler tests for determining fatigue strength are not directly applicable, because the failure mechanism for plastics is completely different from that for metals. Plastics have no slip planes, and strain hardening by microplastic local deformation does not occur. However, the following should be borne in mind:
  - Semi-crystalline plastics can heat up under higher frequency oscillating loads, causing the Young's modulus and yield strength to decrease.
  - In fiber-reinforced plastics, cyclic loading can cause the fibers to detach from the plastic. After many cycles, the component may eventually fail. In this case, Wöhler tests may be appropriate.

2 The Injection Molding Manufacturing Process

The manufacturing processes for plastic components are largely primary shaping processes, i.e. the components are molded directly from the melt. Of these processes, injection molding enjoys the greatest importance. It is estimated that extrusion processes for semi-finished products (profiles, pipes, films, sheets, etc.) consume more plastics in terms of volume. But in terms of the variety of plastic products, injection molding is more significant.

Of the various primary molding processes, injection molding is the most significant

The process is briefly explained below to provide the designer with the basic understanding he/she needs.

2.1 The Process and What the Designer Should Know

Injection molded parts are produced in a cyclical process (Figure 2.1). The plastic is melted in an injection unit and metered in the quantity required for one shot. This melt is injected into a mold cavity at pressures of up to 2000 bar.

General process parameters

A very high injection pressure is present only in the immediate nozzle area of the injection unit. From the injection point to the end of the flow path, the pressures drop overall. At a rough estimate, an average pressure of approx. 300 bar acts on the component's surface.

Compared with the melt temperature, the mold is cold. This allows the injected plastic to solidify and be demolded as a component after the mold opens. Because the volume of the plastic decreases during cooling, further plastic melt must be pressed into the cavity after injection and during cooling. This holding pressure is only effective up to a certain cooling temperature, i.e. the plastic will continue to shrink during the cooling phase and will eventually occupy a smaller volume than the cavity.

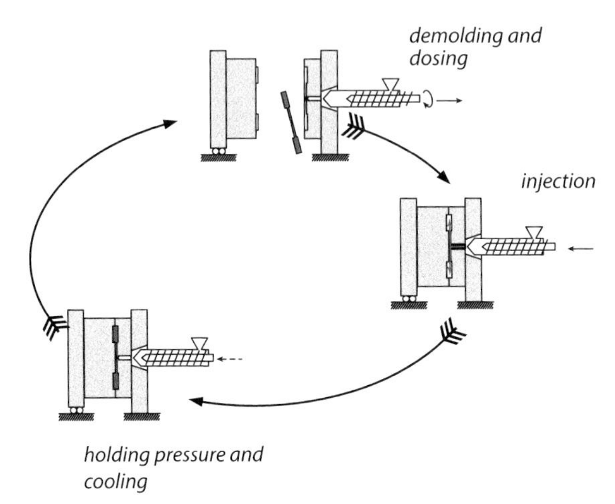

Figure 2.1
Injection molding as a cyclical process and the essential process steps

Important general information for the designer

For the designer, there are four essential pieces of information to know in the first instance:

- Flow path lengths and thus the part size are limited or multiple gating points may be required
- Component area determines the clamping force and thus the machine size
- Wall thicknesses determine the cycle speed and thus the costs
- Shrinkage changes the component size

2.1.1 Flow Path Lengths Are Limited

Filling pressure limits the flow path length

The flowability of the plastic makes these high pressures necessary and limits the flow path lengths that are possible with an injection molding machine of a given performance class. The physical basis for this is summarized in the Hagen-Poiseuille equation:

$$\Delta p = \mathrm{K}\frac{\eta L v}{G^2} \tag{2.1}$$

where Δp is the difference in pressure between the nozzle and the flow front that is necessary to make a melt of viscosity η flow with a velocity v over a distance L and a flow cross-section G (Figure 2.2; K stands for a constant proportionality factor). The viscosity is a value that describes the flowability of the melt. High values mean poor flowability and thus high filling pressures. G is the characteristic thickness, i.e. the diameter in the case of a cylindrical cross-section and the predominant thickness in the case of a flat component.

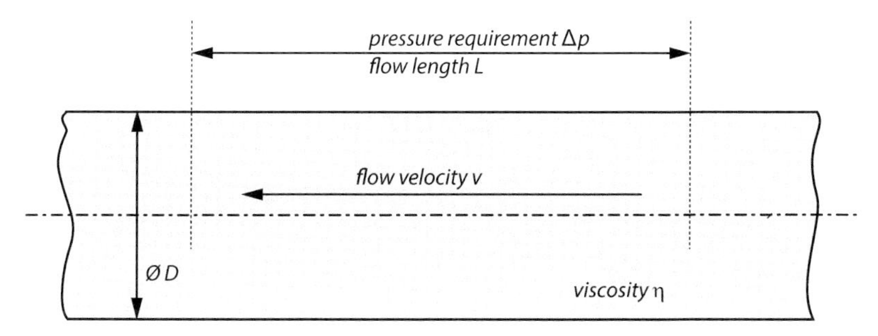

Figure 2.2 Pressure loss or pressure requirement for the flow of a plastic melt

Calculation of the flow path lengths cannot be reliably carried out with the very simplified Hagen-Poiseuille equation, because the viscosity is very strongly influenced by the temperature, among other things. However, the viscosity cannot be represented by a single number, because the melt is injected into a relatively cold mold and so its temperature varies with its distance from the cavity wall. If the flow path lengths are to be determined reliably, a rheological simulation is necessary (see Chapter 4 "Simulation"). These simulations allow the injection process to be estimated very reliably.

The results of a flow spiral test can also be used to arrive at estimates. These are strip-shaped components that have very long flow paths (Figure 2.3). Depending on the flowability of a plastic, different flow path lengths can be achieved under predetermined injection pressures.

Maximum feasible flow path lengths can be estimated with the flow spiral

Figure 2.3 Flow spiral for measuring the flow path length

Figure 2.3 shows the usual flow path lengths resulting from flow spiral tests. According to this, the ratios of flow path length to wall thickness are not greater than 150. Special materials that flow particularly easily are used in packaging applications. In that event, flow path length ratios of up to 300 are possible.

As components become larger, it is inevitable that several injection points will have to be used. A sheet approx. 300 mm wide, 600 mm long, and 2 mm thick could still be filled via two injection points. However, a weld line is inevitable where the melt flows merge.

2.1.2 Molded Part Area Determines Machine Size

Clamping force, molded part area, projected area, and machine size

The two mold halves are pressed together by the clamping unit of the injection molding machine. The clamping force of the machine must be greater than the force acting in the cavity in the direction of the opening axis of the machine. This force results from the projected area of the part and the average cavity pressure. The average cavity pressure is assumed to be roughly 300 bar.

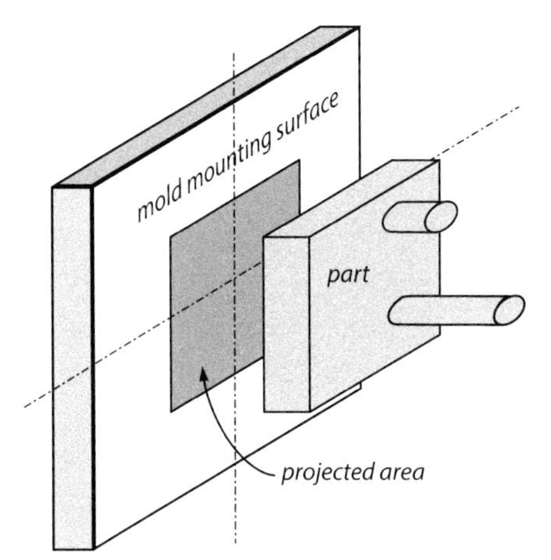

Figure 2.4
Projected area of an injection molded part

Estimation of clamping force

A rule of thumb is sufficient for estimating the clamping force:

$$F_{clamp} = 3\,A_{proj}\,\frac{\text{kN}}{\text{cm}^2} \tag{2.2}$$

This calculates the clamping force in kN for an assumed average pressure of 300 bar acting over a projected area A_{proj} in cm^2. For a component with an area of 100 cm^2, a clamping force of at least 300 kN, or 30 t, would therefore be required.

The machine size has an influence on the manufacturing costs

In this way, the size of the molded part determines the size of the machine required and thus also the manufacturing costs. The machine size is determined by the clamping unit; a range of models exist in which the clamping surface and clamping force increase uniformly. However, there are a few exceptions where the required machine size cannot be determined directly via the projected area. For example, flat screen frames have a large dimension but only a small projected area.

In this case, the machine size is determined not by the clamping force, but by the size of the mold that is to be installed in the machine.

2.1.3 Wall Thicknesses Determine the Cooling Time

Estimating the cooling time and cycle time from the wall thickness

After movement of the clamping unit for opening, demolding, and closing the mold, the molten plastic is injected into the cavity in a cycle and under enormous pressure. The cooling process begins as soon as the melt makes first contact, even though the cooling time used for the production parameter starts after the holding pressure phase. The cooling time that elapses between the end of the holding pressure and opening of the mold is available for melting and metering the melt again for the next cycle (Figure 2.5).

The cooling time itself can be estimated by means of a temperature calculation. More specifically, an average demolding temperature is selected at which the plastic component has solidified sufficiently and can be demolded. By average temperature is meant that the component is quite cold on its surface, but is still very hot inside.

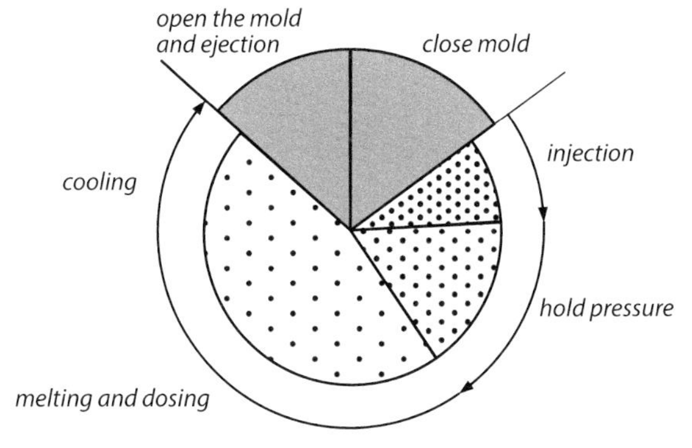

Figure 2.5
Timing of the main process steps

Given the usual thermal conductivity values for plastics, and the melt, mold and demolding temperatures, the cooling time can be estimated as follows:

$$t_{\text{cooling}} = s^2 \text{ s/mm}^2 \qquad (2.3)$$

where s is the predominant wall thickness in mm. The cooling time t_{cool}, is about half as long as the total cycle time in very many applications. Thus, a part with a wall thickness of 2 mm would have a cooling time of about 4 s and a total cycle time of 8 s. The wall thickness has a very large influence on the cycle time, which is why many plastic components tend to be thin-walled compared with other materials and have wall thicknesses that are as uniform as possible.

2.1.4 Plastic Shrinks as It Cools

The high coefficient of thermal expansion of plastics results in high shrinkage

Most materials have a positive coefficient of linear thermal expansion, i.e. they expand when heated and contract again when cooled. For plastics, this characteristic value is about 10 times that of steels. The significance of this for the designer is twofold:

- For components of an assembly made of plastics and metals, the strong expansion must be taken into account, since the plastic parts will come under stress when the temperature changes. It may be possible to achieve stress-free expansion of the plastic components through a clever choice of fixed points.
- To produce the plastic parts, it must be remembered that the hot plastic contracts in the considerably colder mold cavity and ultimately occupies a smaller volume than the cavity of the mold. To achieve a desired dimension, the cavity must therefore be larger by this shrinkage dimension.

Thermal expansion is directly related to a change in density, i.e. the mass per volume in g/cm^3. For processing, the reciprocal of the density is important, and is called the specific volume (cm^3/g).

Increasing the pressure within the process to compensate for shrinkage

Figure 2.6 illustrates the change in specific volume. The melt is strongly compressed during injection. For a given cavity volume, further plastic is therefore pressed into the mold at high pressure, whereby the density increases and the specific volume decreases (1 → 2). During cooling, the plastic shrinks and further plastic can be pressed into the cavity while high pressure is maintained (2 → 3). Above a certain temperature, the plastic has so little flowability that no further plastic can enter the cavity, even under high pressure. Therefore, during further cooling, the specific volume does not change (3 → 4) while the pressure steadily decreases. When the pressure in the cavity reaches ambient pressure (1 bar, point 4), the plastic continues to contract and detach from the cavity wall while the mass remains constant (4 → 5). When the plastic has cooled sufficiently and is stiff enough, the part can be demolded (5). After demolding, the volume continues to change until the component finally reaches ambient temperature completely.

Effects of plastics shrinkage on the design

For the designer, this has the following implications:

- The mold that must be built to produce the component will have to be larger than the component. In many cases, this is the mold maker's first consideration.
- The shrinkage behavior of plastics is not uniform in all spatial directions. Particularly in the case of fiber-reinforced components, it must be noted that the orientation of the reinforcing fibers strongly influences the shrinkage behavior. Glass fibers, for example, undergo much smaller shrinkage than plastic, and so a component shrinks less in the longitudinal direction of the fibers than in the transverse direction.

Shrinkage behavior can be predicted with simulation software. Therefore, it is very useful in component development to simulate the manufacturing process in parallel with the CAD design and to consider possible changes in shape at the design phase.

Simulating shrinkage behavior in the design saves time-consuming rework after initial sampling

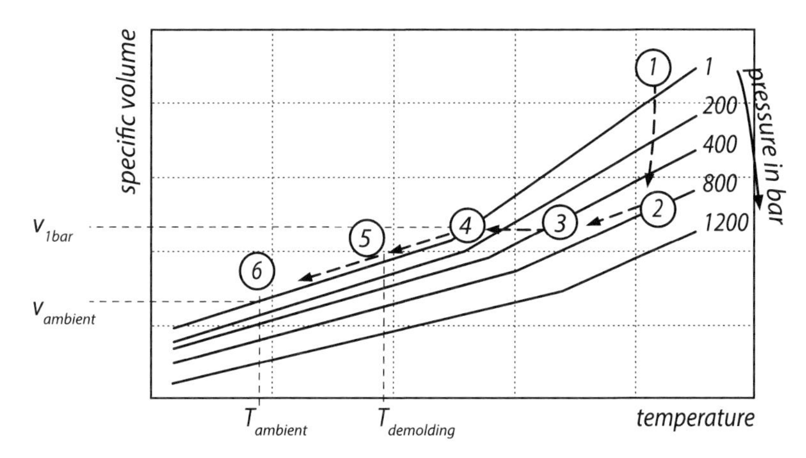

Figure 2.6 Volume change of a plastic during the injection molding process
1: Melt largely pressureless
2: Melt under the influence of the injection pressure
3: State after end of effective holding pressure
4: State when ambient pressure is reached
5: State during demolding
6: State after complete cooling

2.2 Influence of the Process on Component Properties

The injection molding manufacturing process has a crucial influence on the properties of the manufactured components. Affected are the:

Effects of the injection molding process on component properties

- *Component surface,* because flow lines – more precisely, weld lines – are formed at the points where two melt streams meet, and the surface quality in terms of gloss or deliberate texturing is reproduced to a greater or lesser extent.
- *Component shape,* because warpage may occur, i.e. the component undergoes unwanted bending, because the melt flow causes orientation. Orientation here refers specifically to the alignment of fibers when fiber-reinforced plastics are used.

2.2.1 Weld Lines, Meld Lines

Weld lines can impair the strength of components

Weld lines (knit lines) can adversely affect both the appearance and the strength of the component. During injection molding, the melt flows around obstacles, such as cores in the mold, which create breakthroughs, holes, or openings in the component wall. These cores divide the melt, with the result that two melt fronts come together again behind the cores and join or weld there.

However, it may also be the case that the melt flows faster in somewhat thicker molded part areas and later meets again with a flow front coming from a second flow direction. When two melt streams meet and continue to flow together, what are often referred to as meld lines arise; these are more visible at the point of meeting but become progressively weaker. In principle, these are also weld lines, because both melt streams must weld together as well as possible to prevent mechanical weakening and impairment of the component's appearance.

Formation of weld lines

At the flow front, hot melt always swells out like a fountain toward the surface from the center of the cross-section. Immediately behind the flow front, the melt meets the cold mold wall, and a solidified surface layer is formed (Figure 2.7). When two melts meet, they must weld together as perfectly as possible.

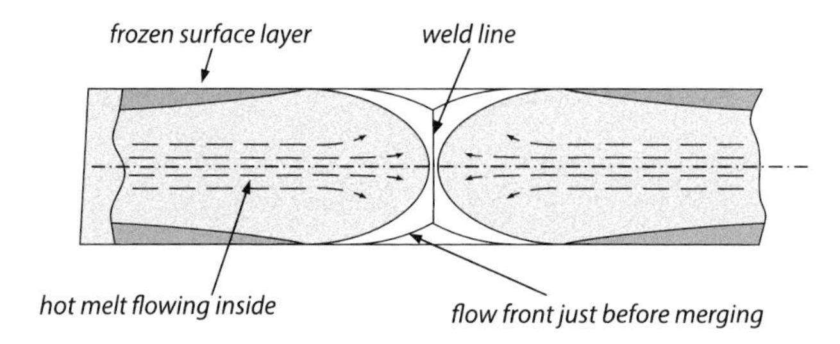

Figure 2.7 Formation of a weld line when two melt flows meet

Quality of the weld line can be controlled in the design

The quality of the weld is directly related to the temperature and time because the molecules coming together need to mix or interlock as well as possible. It quickly becomes apparent that these conditions are more favorable in the center of the cross-section than at the relatively cold mold itself. In many cases, it can be observed during production that the weld line is still barely visible, if at all, immediately after demolding but becomes visible as a fine hairline crack during further cooling. This is because the plastic, which continues to cool, contracts, and thus generates stresses that the weld line strength in the edge area cannot withstand.

Figure 2.8 also shows that the flow direction of the melt directly in the weld line is not longitudinal, but transverse to the component axis. Particularly in the case of fiber-reinforced materials, this leads to considerable weakening of the component strength.

The weld line, and especially its visibility, is not constant behind a obstruction. Immediately behind the breakthrough, the melt meets head-on (Figure 2.8). The angle at which the flow front converges becomes smaller with increase in distance from the breakthrough. The smaller this angle, the more likely it is that melt will continue to flow into the component at these points. This further flow movement generates frictional heat inside the component, which ensures better welding.

Change of flow direction within the weld line

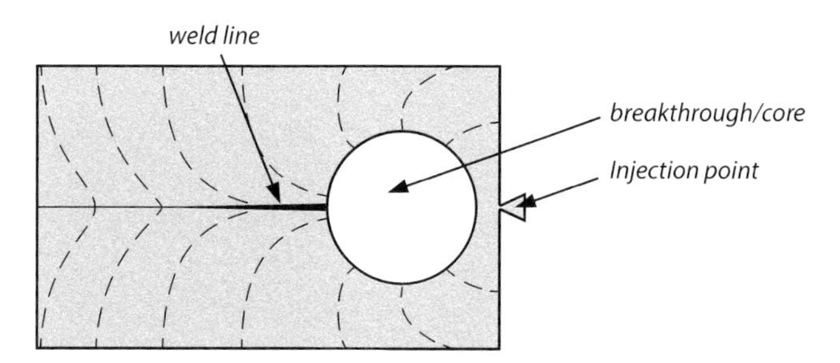

Figure 2.8 Visibility of a weld line decreases with increase in distance from the origin

Overall, it can be stated that weld lines degrade surface quality and are a problem as regards strength. This does not mean that designers should therefore dispense with openings, but they should be aware of the location of weld lines. Computer programs are available that can predict the filling pattern of a cavity very well. With this support, the designer can determine better locations for injection points, for example.

Whether there should be weld lines and where they will occur should be factored into the design of the mold from the beginning

The designer should also bear in mind that weld lines can be a greater problem at the end of the flow path, because less frictional energy will occur here.

2.2.2 Surface Quality

The surfaces of plastic components often have defined textures. These may be erosion structures or special patterns that have been etched into the mold surface and transferred to the component surface during the injection molding process.

The hotter the mold, the duller the component surface

The plastic melt is relatively viscous and therefore does not immediately reach every small depression in the mold surface (Figure 2.9). It is not until the end of injection and the end of the flow path that a high overall pressure is built up, which forces the plastic surface into the smallest depressions. Even though the melt gives off less heat on a rough mold surface because of the smaller contact area, a solidified edge layer forms, which makes the melt even more viscous.

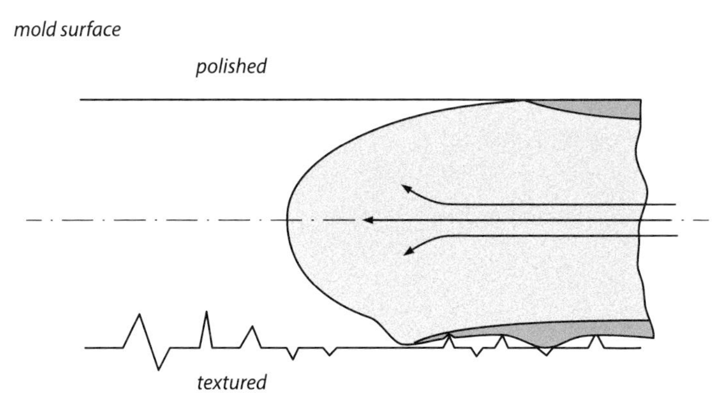

Figure 2.9 Molding of mold surface texture in relation to the mold temperature

The surface quality is strongly influenced by the mold temperature. Especially with rough mold surfaces, the hotter the temperature of the cavity surface, the more matte the part appears (Figure 2.10), because the incident light is scattered more and thus less light is reflected in the direction of the observer.

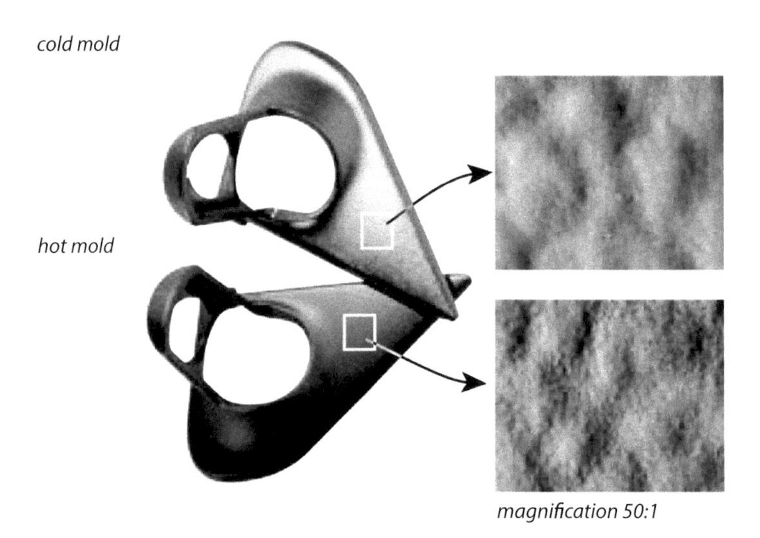

Figure 2.10 Component surface in relation to mold temperature [image source: BASF]

High mold surface temperatures lead to good reproduction of the surface

For a good impression, the surface temperatures of the mold should therefore be hot and the injection times as short as possible. On one hand, the designer could plan shorter flow path lengths over several injection points, but this would inevitably produce additional weld lines. On the other, he/she could demand special technologies directly from the mold maker that ensure the mold temperatures are as high as possible for the injection process (see Section 4.5).

■ 2.3 Fiber Orientations Influence the Component Dimensions

The dimensions of plastic components often do not correspond exactly to those of the mold. In addition to shrinkage (see Section 2.1.4), the reason is internal stresses which can lead to warpage. Whether the internal stresses cause warpage ultimately depends on the stiffness of the component and thus on the shape and temperature. At high temperatures, Young's modulus decreases and then a component can warp. This can also happen sometime after production.

Influences of shrinkage and internal stresses on component dimensions

The causes of internal stresses are varied:

- *Cooling stresses:* If the two halves of the mold are at different temperatures, the surfaces of a component will not have the same temperature after demolding. During further cooling outside the mold, the warmer side may shrink more, with the result that the component deforms toward the warmer side.
- *Crystallization differences:* Some plastics can partially crystallize during cooling. The degree of crystallization is influenced by the cooling rate. The faster a component cools, the lower the degree of crystallization. The degree of crystallization itself influences the density and thus also the component dimensions. Different temperatures within a mold can thus cause distortion.
- *Orientations* are alignments. The flow of the melt can align both the molecules themselves and any fibers that may be present.

Most plastic components are "inherently rigid", i.e. edges and ribs stiffen the component to such an extent that the above-mentioned causes of warpage largely have no effect. Only fiber-reinforced plastics require extreme caution.

Design elements such as edges and ribs counteract warpage

Compared with the plastic melt, the fibers have a much higher melting point and undergo considerably less thermal shrinkage. Data sheets, therefore, always state values for longitudinal and transverse shrinkage. Figure 2.11 clearly shows that the molecules can undergo more shrinkage transverse to the orientation of the fibers. For good strength, the molecules must adhere well to the fibers. This more or less fixes the position of the molecules during cooling. In places where there are no fibers, the molecules can shrink or be stretched, depending on the shrinkage constraint.

For fiber-reinforced plastics, a distinction must be made between longitudinal and transverse shrinkage as regards warpage

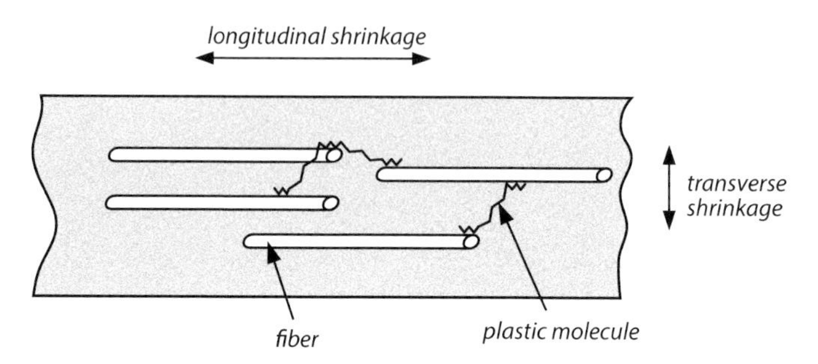

Figure 2.11 Due to obstruction by the fibers, the plastic will undergo only slight longitudinal shrinkage during cooling

The orientation of the fibers is influenced by the injection molding process (Figure 2.12). The melt flowing through an injection point into a cavity spreads out radially around the injection point in the cavity. The radius increases steadily, resulting in elongation transverse to the direction of flow. This elongation causes the incoming fibers to rotate transverse to the flow direction. Immediately upon wall contact, the melt solidifies and stops flowing. Closer to the center of the flow, the melt is moving faster. Between the melt particles of different velocities, shear flow occurs. Here, the fibers are aligned in the longitudinal direction. At the flow front, the melt swells out like a fountain and the fibers are again deflected.

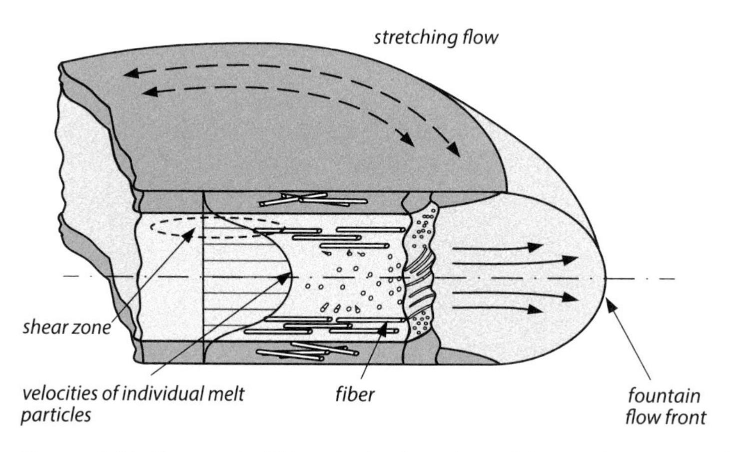

Figure 2.12 Alignment of fibers in the flow channel due to shear, stretching, and fountain flow

Without simulations, the designer has no way to determine the orientation of the fibers in advance

Zones of different fiber orientation are formed across the cross-section. In the center of the cross-section, the fibers are predominantly transverse to the direction of flow; right at the edge and at the surface, the fibers are predominantly unoriented, i.e. they are evenly distributed in all directions. Between these two zones is the shear zone in which the fibers are largely oriented in the direction of flow. The designer has no chance of knowing the fiber orientation unless he/she simulates the injection process with a computer program. Such simulations provide clues as to which molding areas have aligned fibers and how and in what way this will affect likely warpage of a part. Armed with this information, he/she can change the position of the gating point and, if necessary, adjust the wall thicknesses so that the part will take on the desired shape.

■ 2.4 Forward-Looking Quality Assurance

Avoidable design errors in injection molding

The causes of injection molding errors do not only lie in production. The following errors can be avoided early in the design phase:

1. Weld lines (see Section 2.2.1)
2. Sink marks
3. Diesel effect
4. Deformation during demolding
5. Incomplete filling
6. Burr formation

2.4.1 Sink Marks

Sink marks manifest themselves as slight depressions on the surface

Material accumulation may occur in certain areas, e.g. rib feet (Figure 2.13). Here, higher volumetric shrinkage occurs, giving rise to either sink marks or shrinkage cavities. The shrinkage cavities are not visible in non-transparent components, but they weaken the structure. A snap-fit hook might break off unexpectedly as a result, for example. Sink marks manifest themselves as slight depressions on the surface.

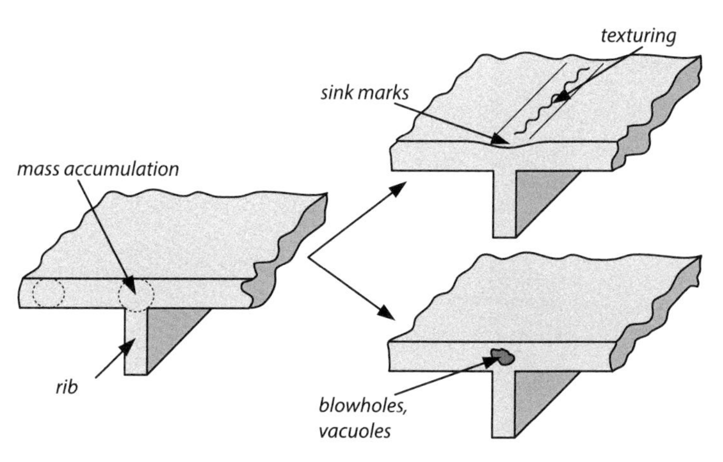

Figure 2.13 Sink marks and voids caused by material accumulation

Design measures for avoiding sink marks

Sink marks can be avoided or at least concealed via the design:

- Reduce material accumulation
- Stiffen the surface with a convex geometry
- Adding texture to the surface

A reduction in material accumulation is possible through slimmer ribs, but this comes at the expense of the desired stiffening, which is often a reason for ribbing in the first place. A convex surface with as small a radius as possible provides initial stiffening for the surface itself. Upon demolding, the component edge zone is colder, while higher temperatures are still present in the interior. Unlike a concave surface (curved inward), a convex surface supports itself, like an eggshell, and can withstand much higher pressures from the outside than from the inside. However, stiffening the surface increases the risk of shrinkage, because shrinkage of the plastic cannot be avoided.

Optical concealment of sink marks through surface texturing

Surface texture can hide sink marks. This works particularly well if the rib path is imitated by, e. g., longitudinal grooves or elevations.

2.4.2 Jetting

Jetting: Uncontrolled injection of the melt through the cavity

The desired type of melt flow is fountain flow. In this case, the melt flows in a laminar fashion between two solidified boundary layers and swells at the flow front from the center of the cross-section toward the cavity surface (see Figure 2.12). Jetting is an unintentional type of flow, where the melt sprays uncontrolled through the cavity (see Figure 2.14). Jetting mainly occurs at cross-sectional constrictions, e. g. in the gate area. With the melt flowing at an overall constant rate, it accelerates substantially on passing through the narrower cross-section and can lose contact with the mold wall.

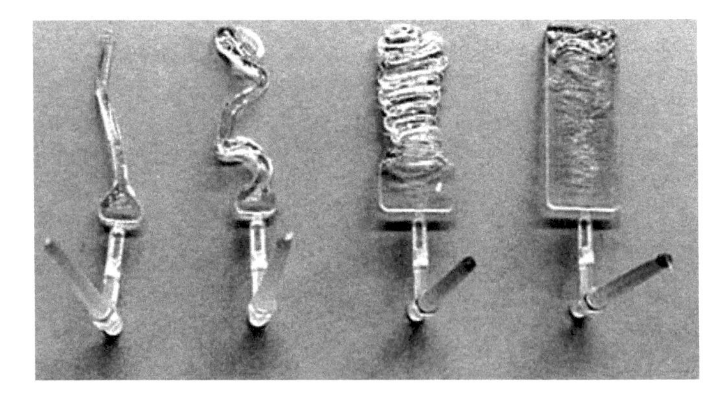

Figure 2.14 Jetting behind a gate; visible due to partial filling

Jetting appears on the surface as an accumulation of many weld lines. It can be prevented by the following (Figure 2.15):

Avoiding jetting

- Having a larger gate, making the sprue point more visible and making it more difficult to separate the part from the sprue system.
- Directing the melt flow against an impact surface, e.g. via a sub gate.
- Creating a movable impact surface, e.g. via a retractable pin directly opposite the gate surface.

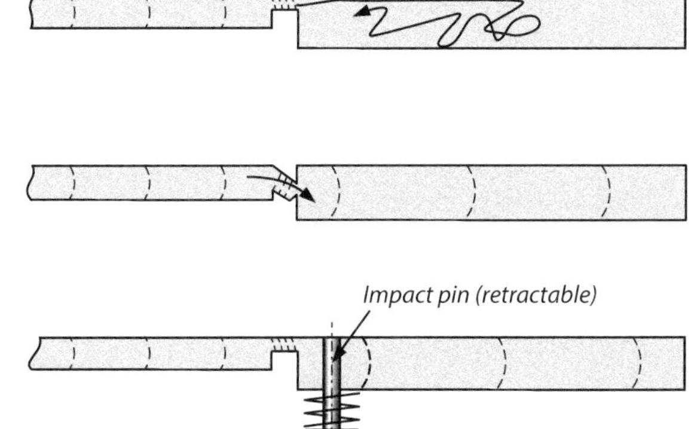

Figure 2.15
Jetting due to accelerated melt flow in the gate area and remedial measures using a sub gate or retractable pin

2.4.3 Diesel Effect

The diesel effect is the self-ignition of a gas mixture due to heating as a result of compression. In injection molding, the air trapped in the mold must escape when the melt fills the cavity. If the air cannot escape, it is compressed and becomes so hot that the plastic is ignited. Localized combustion occurs where the air could not escape.

Strong compression of trapped air inside the mold leads to localized burns on the component

The problem areas are rooted in the design and execution of the mold:

- If several melt fronts do not meet at the component edge, the air can become trapped because it is not discharged via the parting line. In that case, the mold maker has to discharge the air via additional mold parting lines, e.g. via inserts or separate pins (see Section 3.1).
- Often, the air cannot escape fast enough across the parting line and users will then solve the problem with a slower injection speed. However, it makes more sense to provide a vent and thus leave the process window as large as possible for the selection of the injection speed. A vent is a small gap less than 0.02 mm high, which is often manually ground into the parting line of one of the mold halves at the problem area during first mold sampling.

Vents allow the air in the mold to escape and counteract the diesel effect

2.4.4 Incomplete Filling, Burr Formation, and Deformation during Demolding

Causes of incomplete filling of the components

Incomplete filling can have two causes:

- Combustion ("dieseling") does not always occur; for example, when very thin areas are filled, the melt there flows more slowly (see Figure 2.16) and the hot air transfers its heat to the mold quickly enough. The temperature is then no longer sufficient to ignite the plastic. The compressed air generates a counterpressure that impedes filling of the cavity area. One remedy here is to provide better venting.
- If the wall thicknesses are very different, the melt will always flow more slowly in the thinner areas. It can then happen that the melt emits too much heat to the mold and is too cold for adequate filling. One remedy is to make design adjustments to the wall thicknesses.

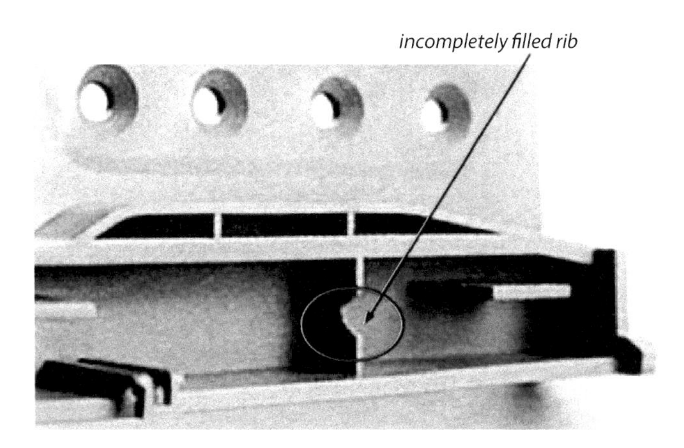

Figure 2.16 Incomplete filling of a rib area due to poor venting

Excessive pressure in the cavity leads to burr formation at the parting line

Flashing is the formation of thin burrs that form on the plastic part around the parting line. The cause is both excessive melt pressure and a poorly inked or damaged parting line that does not seal tightly. Excessive pressure occurs when the clamping force of the machine is too low in relation to the pressure in the cavity and the component surface. This error is due to the design than to the choice of a machine clamping force that is too low.

Notes on the avoidance of burr formation in the design phase

- Gaps in the mold should always be smaller than 0.03 mm so that the hot plastic melt cannot penetrate under high pressure. This applies primarily to mold construction, where all mold inserts must be ground accordingly and the mating mold halves must touch cleanly, i.e. make full-face contact.

- If metal inserts are overmolded, there is a great risk of flashing (Figure 2.17). In many cases, the tolerances of the inserts are not so tight that the specified gap dimension can be maintained. Either the gap dimension is too large, in which case flashing occurs, or the gap dimension is too small, in which case there is a risk of mold damage. Here, too, flash formation can be counteracted with flexible seals in the mold. One possible sealing material is a thermoset semi-finished product that forms the edge of the cavity.

Figure 2.17
Burrs in the overmolding area of a metal insert [image source: Hasco]

Uneven demolding can lead to permanent deformation of the component

For reasons of fast cycles, demolding almost always takes place at relatively high temperatures. In that event, the stiffness of the plastic is still quite low. If too few ejector pins have been set or if the draft angles are too small, mechanical and permanent deformation of the component may occur during demolding.

2.5 Special Injection Molding Techniques

Special techniques extend the range of injection molding

Some extended manufacturing processes are available to the designer. In the standard process, the geometric possibilities are limited to more or less thin-walled, shell-shaped components (Figure 2.18). The choice of extended process is restricted by the design specifics. For example, injection molding of melts containing blowing agent is not discussed below, because the resulting effects on the design are only slight. A blowing agent makes it possible to achieve greater wall thicknesses and significant differences in wall thickness, because it generates cavity pressure independently of the machine's holding pressure.

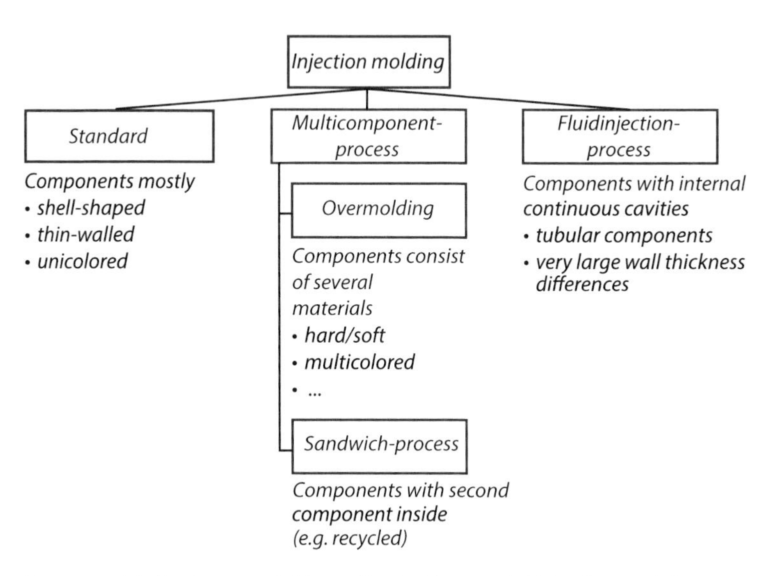

Figure 2.18 Design possibilities afforded by special injection molding processes

2.5.1 Multi-Component Technology

Multi-component technology includes sandwich molding and overmolding

In multi-component technology, several plastic melts are processed into one component in one process. A distinction is made between *sandwich molding* and *overmolding*. In sandwich molding, two melts are injected into the cavity at different times so that the melt flowing in later is hidden inside the component. The proportions of the two components depend strongly on the filling pattern.

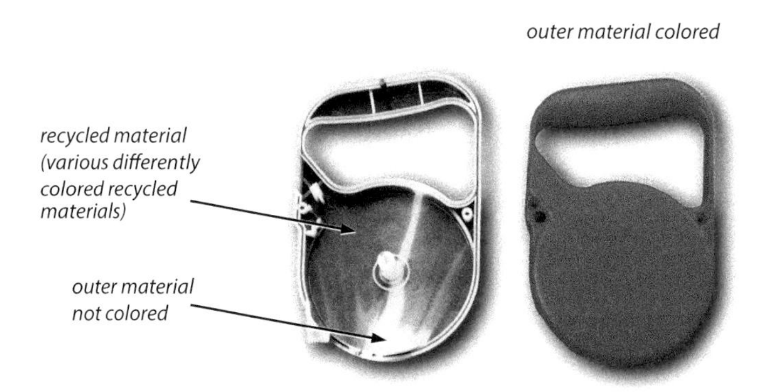

Figure 2.19 A sandwich component often has a non-visible second component inside the component

In sandwich molding, the construction of the components is analogous to that of standard components, the arrangement of the materials (visible outer skin, non-visible material inside) being effected via the process control. This is different for the overmolding process. Here, the design determines the arrangement of the different materials so that, for example, different soft surfaces or lettering can be reproduced. It is also possible to produce components with articulated joints.

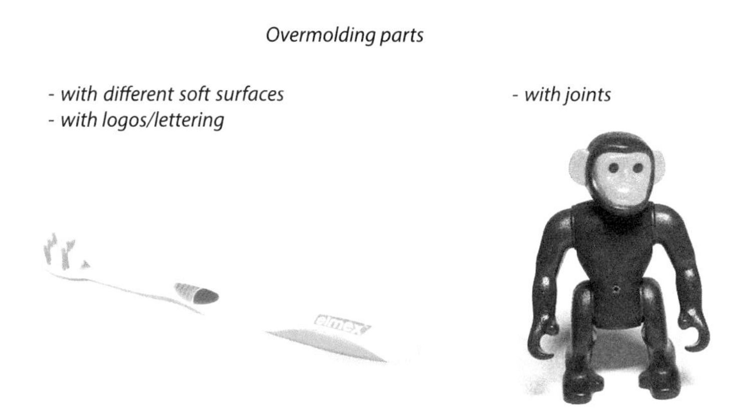

Figure 2.20 Overmolded parts

In principle, overmolding can also be used to produce thick-walled components from just a single plastic component. Dividing the total thickness into several thinner layers which are injected one after the other allows the cycle time to be reduced considerably, because each layer can be cooled from at least one side. The cooled side of the preform is melted a little by the second melt and combines completely with it. This is a good way to produce thick-walled optical lenses.

2.5.1.1 General Procedure

A previously produced basic body or preform is partially overmolded with a second plastic in a further process step. In principle, two independent machines can be used, with a preform being produced on the first and overmolded with a further material in the second. The disadvantage here is the handling of the preform, which must be inserted very precisely into the second mold. For continuous mass production, therefore, machines and molds are used that enable two materials to be processed simultaneously.

In overmolding, a preform is overmolded with a second plastic material

The transfer technique is like the two-machine technique, but both processes take place on one machine. Such machines usually have two injection units, so that two different materials can be used. The molds have two cavity areas. In the first area, a preform is produced, which, in the second process step, is overmolded in the second area. Because the preform is only transferred within the mold, this can be done with precision and repeatedly by a handling system (Figure 2.21). The advantage here is that positioning of the component in the gripper is simple, because the

The challenge of the transfer technique is precision transfer of the preform from one cavity to the other

gripper always removes the component from the same position, and so the component is precisely oriented in space. This is the only way for a gripper to insert the preform into the second mold area without damage. If two independent machines are used, the preform from the first process must first be precisely positioned in a separate setdown area for pickup by a robot.

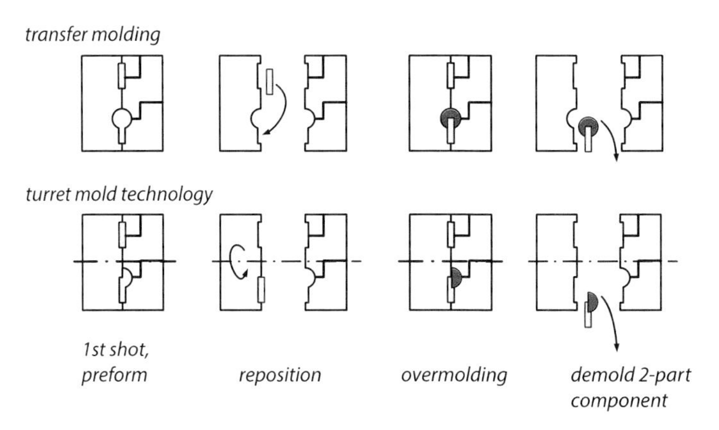

Figure 2.21 In overmolding, a layer of material is molded over a previously manufactured plastic component

Turntable molding bypasses transfer of the preform

In turntable molding, part of the mold is rotated and the preform is thus moved to the next position. Most components are produced in this way, because the rotary movement can be very rapid and the preform does not leave the mold cavity completely. On one hand, this means that it cannot be damaged during reinsertion. In addition, the part is more dimensionally stable if it is not removed from the mold core, which is used to transport it from one area to the next in the mold. On the other, the disadvantage of this variant is that the preform can only be overmolded from the side facing the nozzle.

2.5.1.2 Molding Techniques

Overmolding can be implemented by three different mold technologies:
- Core back
- Transfer
- Turntable

Basically, three different mold technologies are employed in overmolding. Except for the core back technology (Figure 2.25), the processes for the preform (1st component) and the finished part (2nd component) take place simultaneously. Finished parts can then be removed with each cycle. In terms of mold technology, the transfer technique (Figure 2.22) is the simplest. The following design aspects must be considered:

- The preform must be able to be picked up safely by a gripping system.

There are several design aspects to be considered in the transfer technique

- The preform leaves the cavity completely for a short time. Depending on the design, it is pulled off a core in the process and can then "snap together", i.e. become slightly smaller in shape.

- For reinsertion into the second cavity, the preform must have suitable mold inclines as far as possible. Even if it has collapsed somewhat due to short-term relaxation and can only be placed on a second core with force, it must be ensured that the closing movement at the latest pushes the preform securely and completely onto the mold core.
- If necessary, it must be ensured that the preform does not simply fall out of the mold again during the closing movement.
- The cavity for the second component only needs to be precisely machined in the area of contact with the preform. Areas that are not overmolded may only have been very roughly machined and leave sufficient space for the preform.

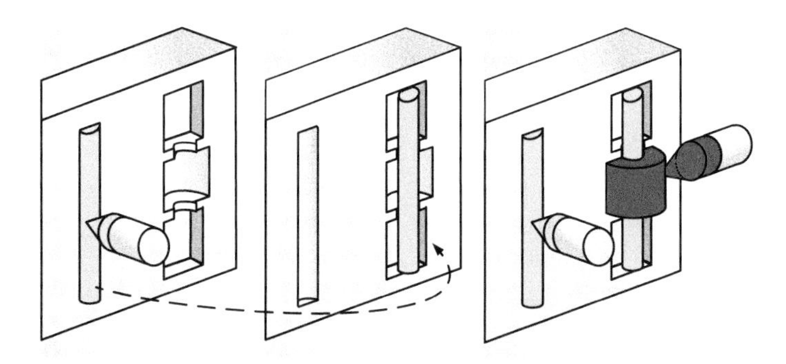

Figure 2.22 Transfer technique during overmolding (the first two pictures show a start-up process)

The turntable technique (Figure 2.23) is used for most overmolding molds. Some molds have an ejector side which is completely rotated through 180°. In this case, the machines usually have a corresponding rotary table on which the mold half is clamped. In many cases, only parts of the ejector side are turned, in which case the rotation technology is integrated into the mold.

Turntable mold technology is the most widely used overmolding process

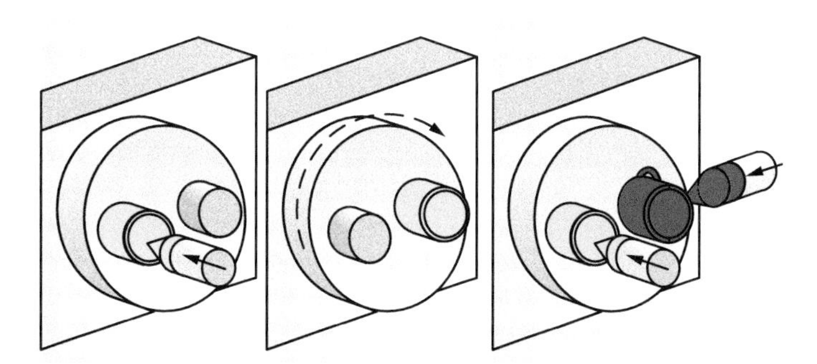

Figure 2.23 Rotary technique used in overmolding (the first two pictures show a start-up process)

For the production of a toy monkey, the head is first injected onto a transport pin (Figure 2.24, *1*). The transport pins and the preforms are lifted out when the mold is open and are moved 120° to the next position via a gear drive (*2*). The very rough rectangular recess for the head can be seen. At this point, only the narrow seal at the neck area is important so that the base body only overmolds the lower joint ball. In the final position *3*, further melt is injected at the joint pins of the torso and at places on the head.

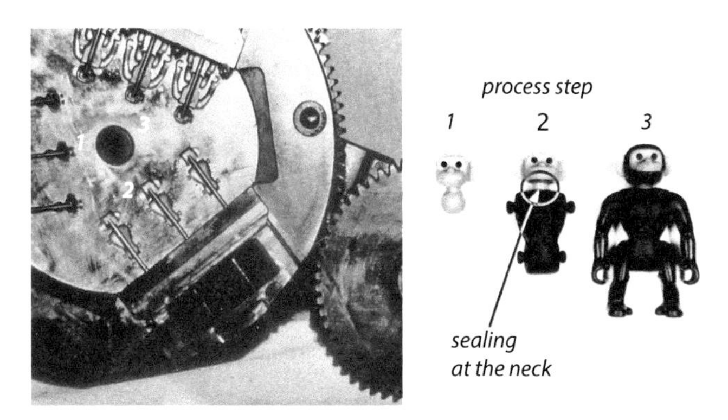

Figure 2.24 Mold for producing a toy monkey, featuring three cavities, each offset at 120°

Turntable molds require several aspects to be considered at the design stage

In terms of design, the following should be noted when using turntable mold technology:

- Where the entire mold half rotates and the preform does not leave the cavity, only the top side can be overmolded. In this case, it must be ensured that the two materials used can be welded together well so that the components do not separate from each other again unintentionally.
- When the preform is moved via an internal rotary drive, it leaves the cavity and can then be overmolded on both sides as in the transfer technique.
- It must be ensured in any event that the preform cannot fall out of the mold during the rotary movement; this is particularly important if it is moved via a rotary drive on the mold.

Core back technology specifications

The core back technique (Figure 2.25) is not very common. Here, the two processes of pre-mold production and molding of the second component take place in succession. Part of the mold cavity can be retracted into the mold, creating space for the second component. This technique is used in the production of seals that are molded directly onto housing frames. With this technique, it is important to note:

- The first component does not necessarily have to be completely cooled down in order for the second process step to begin. The mold remains constantly closed for the core back movement, so that a preform that is not yet stiff cannot deform.

The advantage of this rapid mode of operation is that melts of high temperature come together in the joining zone and can thus bond better.

- Core lifting in the case of largely planar components is only possible in a single plane.

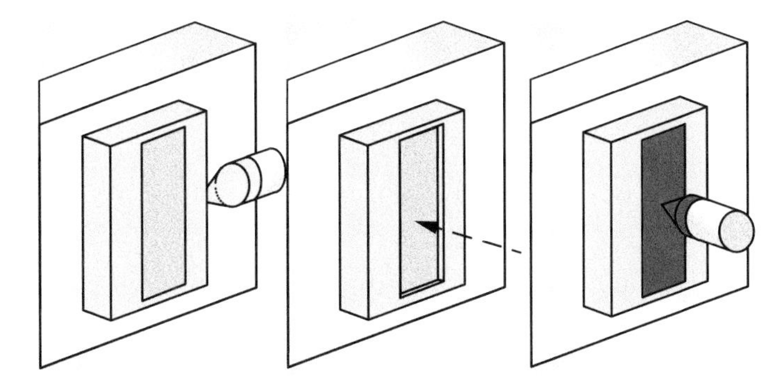

Figure 2.25 Core back technology

2.5.1.3 Component Design

Mold technology influences component design

The part shape depends on the mold technology, e.g. not all of the variants shown in Section 2.5.1.2 allow overmolding of the preform from the front and back sides. In addition, the following questions arise for the designer:

- How to design the joining zone between the components?
- How to ensure the sealing of the cavity for the second component so that the second component does not cover unwanted areas of the preform?
- How to proceed to ensure that moving elements move as smoothly as possible?

Mating Zone (Material Selection)

Knowledge of possible materials that can be welded together is essential for the designer

Depending on the weldability of the materials used, the joining zone between the components has to be designed. In overmolding, a hot second melt is injected onto an already cooled preform. To achieve good joint strength, the second melt must be hot enough to melt the surface of the preform to such an extent that the two melts can weld together. The prerequisite for this is basic weldability. For this purpose, tables exist that show the possible weldability for material groups. The symmetrical overview (Figure 2.26) distinguishes between coextrusion and injection molding. In injection molding, unlike extrusion, two melts do not arrive together, but one melt meets material that has already cooled down. Therefore, less suitable materials can be found here, because the process conditions are less favorable for bonding. Many plastics contain various additives and so the expertise of material suppliers should be sought in addition to standard statements. Additives also include adhesives that are used specifically for certain material pairings.

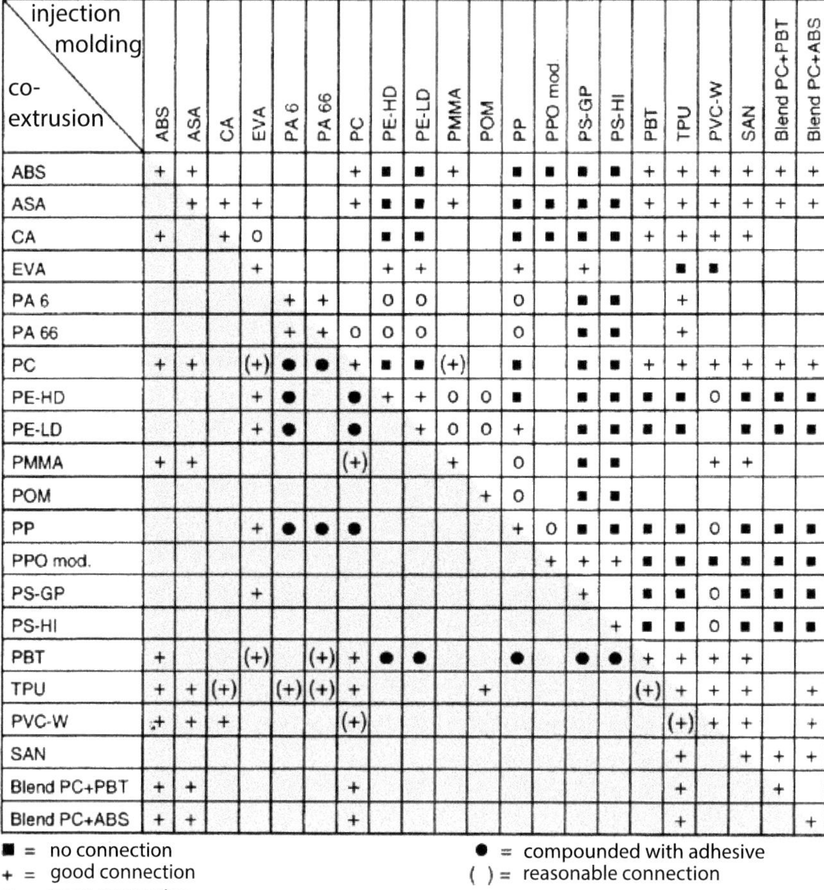

co-extrusion \ injection molding	ABS	ASA	CA	EVA	PA 6	PA 66	PC	PE-HD	PE-LD	PMMA	POM	PP	PPO mod	PS-GP	PS-HI	PBT	TPU	PVC-W	SAN	Blend PC+PBT	Blend PC+ABS
ABS	+	+					+	■	■	+		■	■	■	■	+	+	+	+	+	+
ASA		+	+	+			+	■	■	+		■	■	■	■	+	+	+	+	+	+
CA	+		+	o				■	■			■	■	■	■	+	+	+	+		
EVA				+				+	+			+		+			■	■			
PA 6					+	+		o	o			o		■	■		+				
PA 66					+	+	o	o	o			o		■	■		+				
PC	+	+		(+)	●	●	+	■	■	(+)		■		■	■	+	+	+	+	+	+
PE-HD				+	●		●	+	+	o	o	■		■	■	■	■	o	■	■	■
PE-LD				+	●		●		+	o	o	+		■	■	■	■		■	■	■
PMMA	+	+					(+)			+		o		■	■			+	+		
POM											+	o		■	■						
PP				+	●	●	●					+	o	■	■	■	■	o	■	■	■
PPO mod.													+	+	+	■	■	■	■	■	■
PS-GP				+										+		■	■	o	■	■	■
PS-HI															+	■	■	o	■	■	■
PBT	+			(+)		(+)	+	●	●			●		●	●	+	+	+	+		
TPU	+	+	(+)		(+)	(+)	+				+					(+)	+	+	+		+
PVC-W	+	+	+				(+)										(+)	+	+		+
SAN																	+		+	+	+
Blend PC+PBT	+	+					+										+			+	
Blend PC+ABS	+	+					+										+				+

■ = no connection
\+ = good connection
o = poor connection
● = compounded with adhesive
( ) = reasonable connection

Figure 2.26 Overview of the weldability of important standard plastics

Welding of plastics

The suitability of materials for a composite can also be tested with a simple welding test. In its simplest form, two sets of pellets are melted on a hot plate and the cooled plate is then tested. This qualitative result quickly reveals whether the two pellets can join as a melt.

For materials that weld well, the mating zone should be as smooth as possible (Figure 2.27). An increased surface area due to roughness or shaving tends to be disadvantageous because it makes the heat exchange area too large in relation to the available heat of the melt. The melt heat is supplied by the volume of the second melt. Thus, if a very thin layer is injected onto a rather thick preform, this melt cools down too quickly and the necessary temperature for the joining process may be lacking.

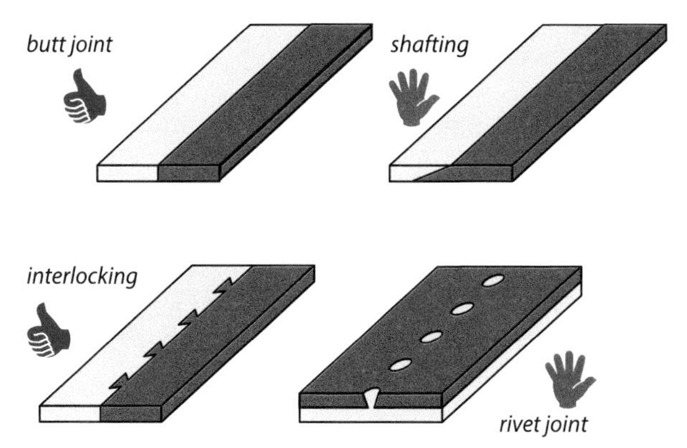

Figure 2.27
Geometrically possible mating zones for two plastics

Mechanical joints between two plastics as an alternative to welding

If the materials cannot be welded securely, positive form locking must be designed. For turntable molds in which the preform does not leave the cavity, the only possible form fit method is the one shown in Figure 2.27, bottom left. A more secure riveted joint, as shown in the bottom right of the picture, is only possible if the preform leaves the cavity and can thus be overmolded from both surfaces. Otherwise, the conical undercut shown cannot be realized.

Sealing of the Second Cavity

Deliberate elastic compression of the preform with the mold to seal the second cavity

The parting line of the second component is formed both by the steel surfaces of the mold lying on top of each other and partially over the preform. The purpose of sealing the parting line of the second cavity, which is formed by mold steel and plastic, is to prevent the second plastic from also overmolding further surfaces of the preform. To this end, the preform is compressed by approx. 0.1 mm in the parting line (Figure 2.28). Due to the considerably lower modulus of elasticity of the plastic compared with that of mold steel, the preform is deformed purely elastically when the mold is closed, even before the mold halves make contact, thus building up a stress which prevents overmolding into the parting line in the event of mold breathing at excessively high cavity pressures. After the clamping force has been relaxed, the preform will spring back to its original dimensions undamaged. This preloading is readily possible for surfaces perpendicular to the clamping movement. The more parallel the surface is with the clamping movement, the lower the sealing effect will be.

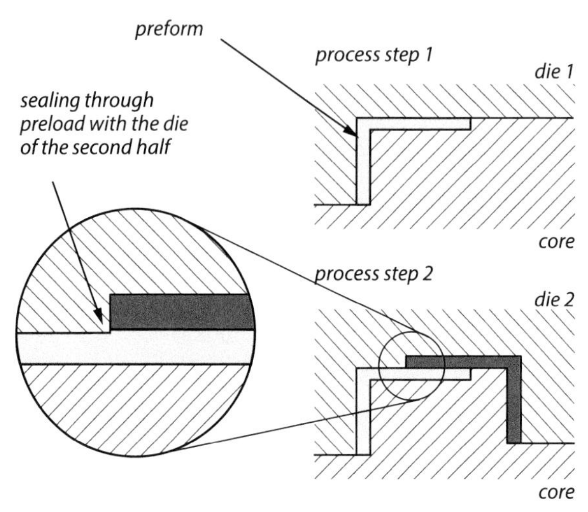

Figure 2.28 Sealing the cavity for the second melt by preloading the preform

High injection pressure poses a challenge when overmolding components over a wide area

In the case of two-dimensional overmolding, the preform may not be able to withstand the high injection pressure and may be pushed away. This danger exists particularly if the preform still has too much residual heat at the time of overmolding and is therefore not stiff enough. To ensure that the boundary line between the two components is exactly straight, it is advisable to form the boundary as a shadow gap (Figure 2.29). In this way, the material of the preform is prevented from escaping the injection pressure at the visible surface by the support of the mold steel.

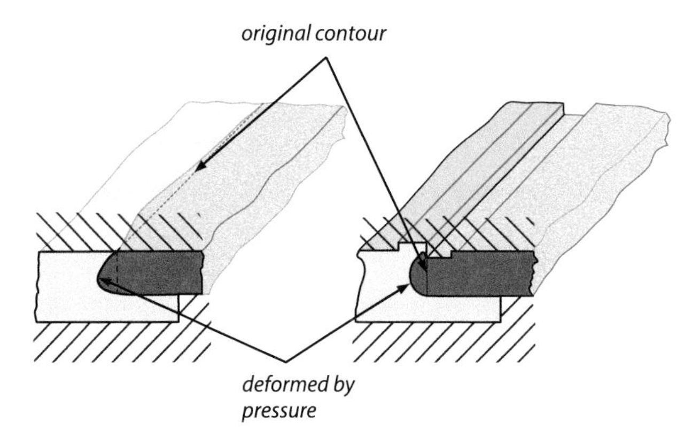

Figure 2.29 Shadow gap for ensuring exactly straight joint boundaries

Sealing for core back technology

In the core back technique, preloading of the preform is only possible by means of an undercut on the moving core (Figure 2.30). The core displaces part of the material of the preform during retraction, and the preform is squeezed.

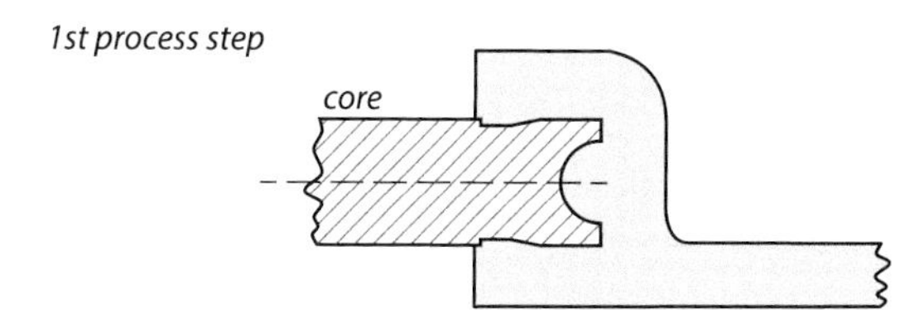

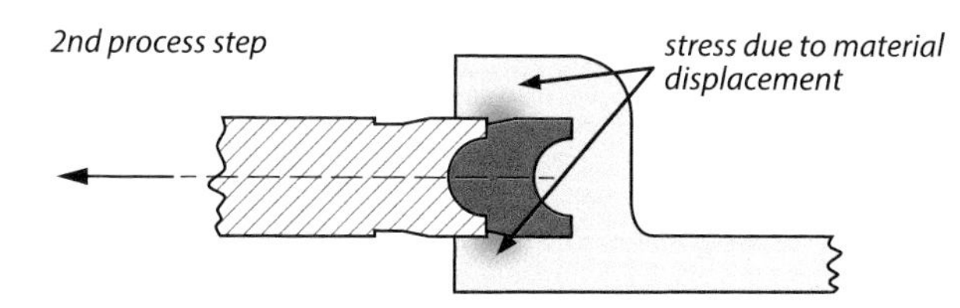

Figure 2.30 Sealing of the cavity for the second process step by means of an undercut on the movable core

Characteristics of overmolding lettering

In the case of multicolored injected lettering and logos (Figure 2.31), the following special features should be noted:

- The finer the lettering, the more attention that has to be paid to the filling process for the second melt. Ideally, a ridge of the preform (e.g. a letter "I") should be reached by the second material from both sides at the same time, such that the injection pressure cannot displace this ridge.
- The entire lettering should be placed on a grid structure of the pre-mold. Then, in the case of the transfer technique, the second melt can flow over the grid from the rear side and the melt can reach the front side of the component through the grid largely at the same time.
- If the rotary technique is used and the preform does not leave the position on the first core, a passage for the second melt may have to be opened via a core back to allow island areas (e.g. the inside of the character "8") to be reached (Figure 2.32).

Figure 2.31 Component structure for lettering (including island areas)

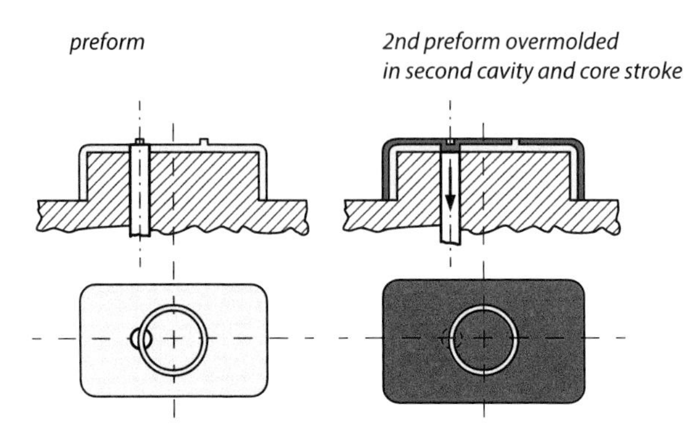

Figure 2.32
Impression of island areas with a combination of rotary molds and core back technology [image source: Arburg]

Components with Movable Joints

Movable joints by the overmolding process

To produce movable joints, plastics that cannot be welded together must be selected wherever possible. Depending on the shape of the preform, a ball joint or a pin hinge can be produced. Advantages over assembled joint connections are:

- Mobility is not dependent on production fluctuations. In the case of assemblies whose individual parts are produced at different times, fitting errors may occur due to tolerance fluctuations. The components are larger or smaller depending on the production conditions. In the case of overmolded joints, if the ball is larger, the ball socket will inevitably also be larger, because the ball itself is the contouring shape for the socket.
- The parts of an assembly can be connected captively and positively. This is not possible in the case of assembled articulated joints, without additional effort.
- Assembly during the overmolding process takes place without the need to pick up and position the individual parts separately. This reduces assembly costs on one hand and reject rates due to faulty assembly on the other.

The force required to move the joint can already be determined in the design phase

The force required for movement is largely dependent on the shrinkage and relaxation behavior of the plastics. Figure 2.33 shows the change in length as a function of temperature. The ball joint of the head is made of PBT and is overmolded at a residual temperature of approx. 100 °C. During complete cooling after demolding of the complete monkey, this ball will shrink even more overall (difference between points 1 and 2) than the body of PA6 molded on top, whose relative change in length at an injection temperature of approx. 270 °C is not shown in the diagram. The relative change in length of the PA6 can be reduced by adding fibers or glass beads, such that the gap between the bead and the body is slightly larger after complete cooling. The good mobility of the head is further improved by pulling the core pin. In Figure 2.33, the head and the ball of the neck joint sit on a pin and are thus moved to their respective positions. When the part is demolded and pulled off the pin, the neck and ball inevitably collapse somewhat, becoming slightly smaller and giving some additional clearance between the torso and ball, so to speak. The arms and legs are made of POM, a material characterized by good sliding behavior. A release analogous to the head is not possible here. If necessary, the mobility could be influenced by changing the shrinkage behavior (e.g. by adding glass spheres).

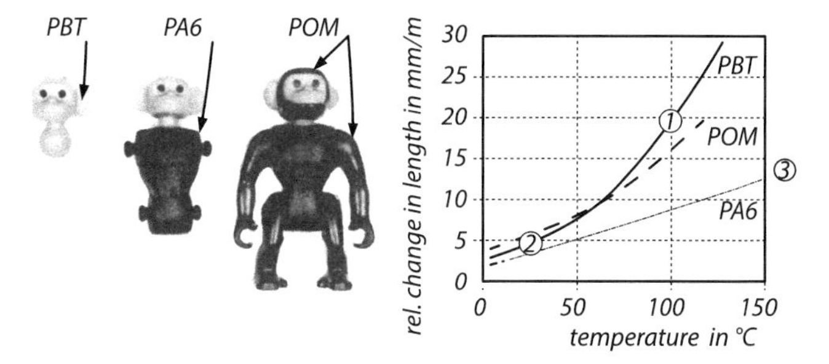

Figure 2.33 Mobility results from the different shrinkage values of the materials used [image source: Ehrenstein, "Mit Kunststoffen konstruieren", Carl Hanser]

2.5.2 Fluid Injection Technology (FIT)

Fluid injection technology for hollow components

Fluids are flowable media, meaning both gases and water in this context. If a fluid is fed into the molten core of a component during injection molding, i.e. between the already solidified surface layers, an internal pressure can be maintained easily and for a long time via this fluid, because its viscosity is considerably lower than that of the plastic. This pressure can just as easily be released again, in which case the outcome is an at least partially hollow component. An additional advantage is that when the fluid is released, the process heat also escapes from the interior of the component.

Process sequence for the fluid injection method

The fluid is fed into the cavity at some time after the melt. The injected melt has already formed a cooled edge layer in contact with the mold, the gas follows the still liquid melt inside the cross-section and pushes it further toward the end of the flow path (Figure 2.34). The fluid has a considerably lower viscosity, so it must be guided in channels. The guide channels are basically thick ribs. Due to the difference in thickness compared with the rest of the component, the fluid will flow in the thicker areas, because the pressure required for flowing is lower there. When the cavity is filled, the plastic increasingly cools and shrinks. As the holding pressure is now taken over by the fluid, the fluid can now flow into the thinner areas. In this case, the fluid no longer has a lateral boundary and thus further fluid spread is uncontrolled. These finger-like shapes are also known as the finger effect.

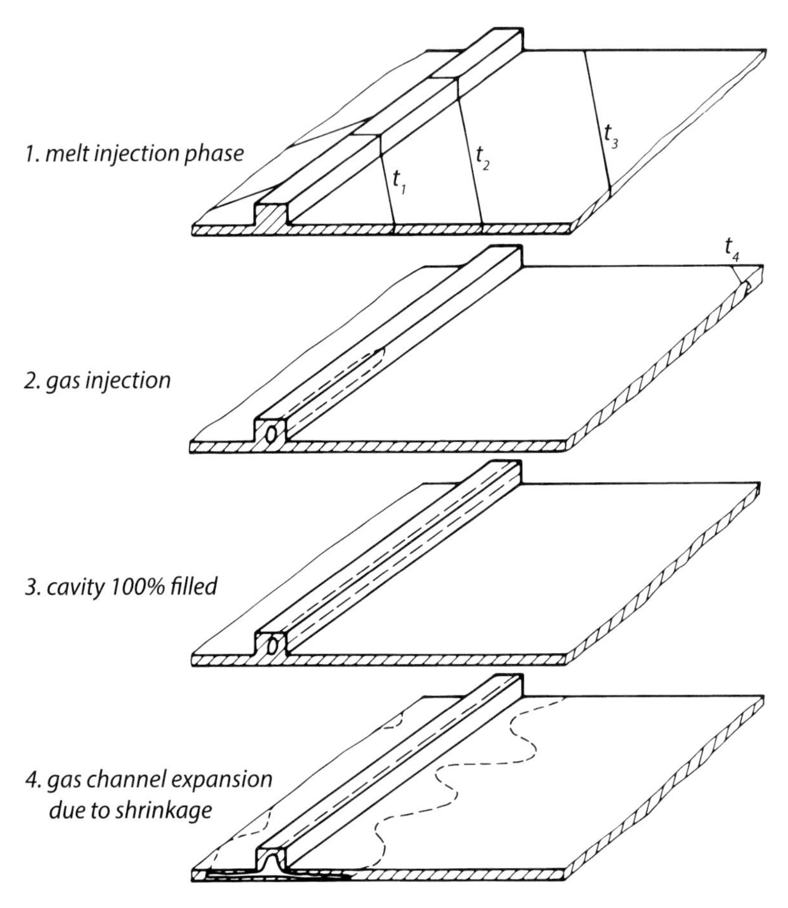

Figure 2.34 Process steps in FIT: A fluid following after the melt creates a hollow space

Two groups of components are mainly manufactured by these processes (Figure 2.35):

- Elongated tubular components
- Components with local thick spots, each of which is hollow

At least one thing they have in common is some large wall thicknesses.

Components with some large and thin wall thicknesses

The components shown are in each case the original version of a metal structure and the plastic variants. Fluid injection components generally do not have any comparative components made of solid plastic, because thick-walled areas cannot be produced in this way in a reasonable cycle time. The main issue here is manufacturability. Advantages such as weight savings or lower manufacturing costs are almost always based only on comparisons with an original metal design.

The designer should be aware that the wall thicknesses in the tubular component area cannot be dimensioned exactly. The expansion of the inner diameter is a result of the process and cannot be selectively modified by the operator of a machine.

In general, it can be stated that most components made by this process technology are design parts. This means that the components do not have ribbed areas but instead have closed surfaces with large wall thicknesses. The reference to "design" parts indicates that technical risks are accepted because a special shape is involved – the main thing is that the product can be manufactured at all. Basically, FIT parts are expensive compared with conventional injection molded parts. This is due on one hand to the additional technical equipment needed and, on the other, to the component thickness and thus long cooling or cycle times.

High financial outlay for the FIT process – therefore more suitable for optical design parts than for technical components

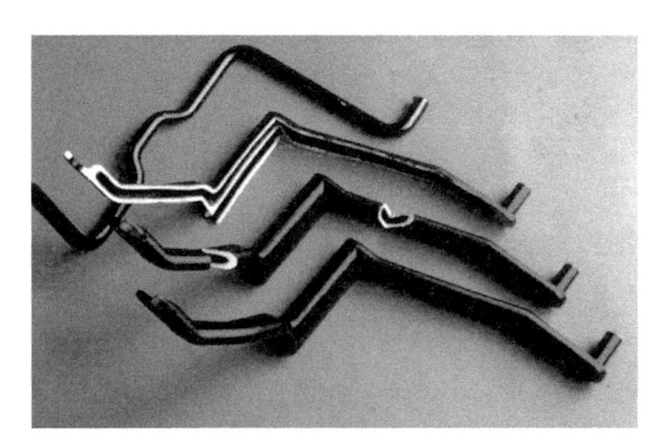

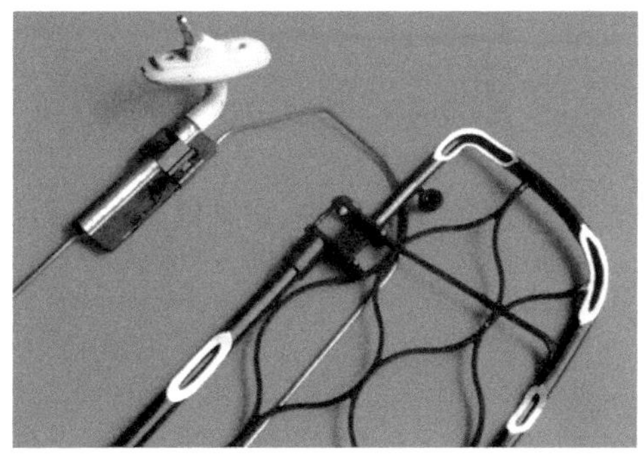

Figure 2.35
Examples of GIT or WIT components.
Above: accelerator pedal linkage;
below: internal construction of a sun visor

2.5.2.1 Processes

The process is distinguished by the fluid employed: GIT uses gas and WIT uses water. GIT mostly employs nitrogen (N_2). It is inert, i.e. it does not burn and can therefore come into contact with the melt even at high temperatures without possible self-ignition. Nitrogen gas has a negligibly low viscosity and can escape from the component at the end of the process without leaving any residue. In comparison, water is disadvantageous because, on one hand, it can lead to corrosion of the mold in the event of possible leakage. On the other, removing the water from the interior of the component is not without problems, and residual moisture may have to be handled or dried on appropriate shelves – a circumstance that may not be easy to solve, depending on the complexity of the cavity formed. In terms of process technology, water is somewhat easier to handle, because it is incompressible

FIT process with water – somewhat more complex than with gas

compared with gas and thus allows more precise control of the cavity formation. Even though the processes are mostly operated with gas, in the following we will generally speak of FIT, because the same basic design rules apply.

The process is divided into three development steps, whereby the process designated FIT-1 below is no longer of major importance but is described here for the sake of completeness.

FIT-1 – Fluid Injection through the Machine Nozzle

The fluid follows the melt

After the melt has been injected into the cavity, the fluid is introduced into the still hot melt via the same nozzle. The prerequisite is that the cavity has only been filled to approx. 60%. Then the fluid can displace the core, which is still molten inside the component, further into the component between the solidified melt walls. The fluid thus follows the melt and pushing it ahead. When the melt has filled the cavity, a hollow space is present inside. Pressure can be exerted on the melt in the cavity via this space if the gas pressure is maintained. This is particularly advantageous for elongated components because the same pressure prevails everywhere in the component perpendicular to the cavity wall. Overall, it is possible to greatly reduce the pressure of the gas after complete filling has been achieved.

Main drawbacks cloud the FIT-1 process

This process variant has some disadvantages:

- Changeover marks almost always occur when the process pressure is switched from melt to gas. Even if the injection pressure is rarely higher than 300 bar for a thick-walled component, deceleration of the melt flow front is hard to avoid. This deceleration causes a solidified skin to form at the flow front, which is then pushed on again in the further course. In the best case, the result is a different gloss at this point or even a perceptible notch, comparable to a poor weld line. Fluid pressures usually do not exceed 200 bar, so the speed of the flow front is slowed by at least 30% at an injection pressure of 300 bar. With higher process pressures, these changeover marks can only be reduced to a very limited extent. Overall, the tightness of the entire system is problematic, especially at pressures above 200 bar.
- If the gas is supplied from the same nozzle as the melt, this has the following disadvantages:
  - The gas must first flow through the gating system. The pressure requirement for this is ultimately missing in the component, at least during the changeover phase.
  - Hot runner molds (see Section 3.4.4) that allow sprueless workflow cannot be used. The gas always flows into the cavity after the melt and thus the gas would inevitably still be in the hot runner after the end of the process. As a result, it

would reach the part surface in the next cycle. A possible solution to this problem would be a combined melt-gas hot runner nozzle. The problem here is that there is very little space available in the area just in front of the cavity, and the nozzle would have to be very small and fine.

- These systems can only be sealed very poorly against leakage, especially in the common melt-gas nozzle. Elastic rubber seals are unsuitable here in the long term. Metal gaskets require much higher fitting tolerances. The possible gas leakages entail higher costs. Usually, nitrogen from cylinder bundles or tanks is used; the gas costs alone can be estimated in simplified terms at around five to ten cents.

FIT-2 – Fluid Injection via Injectors Located in the Mold

The FIT-2 process – fluid injection via special injectors

Injector technology offers a solution to the sealing problem, because the fluid is introduced on the somewhat colder side of the machine-mold system via separate injectors independent of the melt nozzle. Here, the cavity is first partially filled with melt and then the fluid is introduced via the injector, which is installed as close as possible to the gate. The fluid penetrates the hot melt between the solidified surface layers and displaces the melt in the direction in which the cavity is not yet filled. Elastic seals can provide a leak-tight seal here. In terms of design, it should be noted that the injectors can be self-sealed via concentric annular grooves (Figure 2.36). When the plastic cools down, it can also float away from the injector, depending on the volume distribution, allowing the fluid to escape from the interior of the component in the direction of the parting line. An annular groove acts like a labyrinth seal; the plastic cannot float away from the injector in an uncontrolled manner.

Apart from the injectors, this process is identical with the FIT-1 process, i. e. after a partial filling of approx. 60%, the fluid displaces the plastic melt inside the molded part and provides the holding pressure.

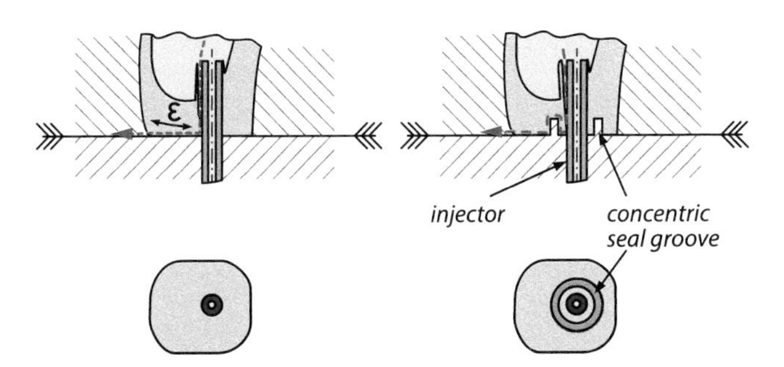

Figure 2.36 Principle behind a fluid injector

The injectors are fixed in the mold and have a diameter of < 5 mm. The melt flows over them and they then protrude into the interior of the component via the rapidly forming solidified surface layer. The injectors vary according to the fluid:

Differences between gas and water injectors

Gas injectors are permanently open and therefore do not require any control. The gas outlet opening is usually formed by an annular gap smaller than 0.03 mm. Plastic cannot flow through gaps of this size. In the simplest case, an injector can be formed from an ejector pin and an associated ejector sleeve. If necessary, the pin can be ground flat over its length just before the opening so that the resistance for the gas is reduced. A fundamental problem is that when the pressure is relieved at the end of the process, not only nitrogen but also decomposition components of the hot melt flow back through the injector. Over time, the gas injectors become clogged and must be removed for maintenance. Systems that can be changed from the cavity side with the mold open are therefore advantageous.

Water injectors require considerably larger opening cross-sections and can therefore be actively closed or opened. The reason lies in the incompressibility of water. While gas can relax once it has reached the component and thus occupy more and more volume, the volume of water must flow in completely through the opening of the injector to form the cavity. To ensure that the formation of the hollow space does not take very long, large cross-sections must therefore be provided.

FIT-3 – Blow-Out Method

The blow-out process combines conventional injection molding with fluid injection technology. After the cavity has been completely filled and a holding pressure applied, hot melt is blown out of thick-walled areas of the component so that these thicker areas become hollow.

Advantages in the FIT-3 process

The advantages of this technique are:

- The components have no changeover markings, because the surface is formed exclusively via a conventional injection phase under a uniform injection speed.
- The finger effect (see Figure 2.41) is avoided, because the melt holding pressure pushes so much melt into the thinner areas that are cooling down that no more fluid can reach them. In addition to surface defects arising from different cooling conditions, the finger effect can also result in a loss of strength.

In the FIT-3 blow-out process, conventional holding pressure can additionally be applied before blow-out

The melt can either be pushed back in the direction of the machine nozzle (melt pushback process) or into an additional space next to the actual cavity (secondary cavity process). Particularly for components that are not only tubular but also have thin-walled areas, conventional holding pressure can be applied before the blow-out process.

The melt back-pressure process exploits the large wall thickness in the fluid channel area. Here, the melt cools down so slowly that it is transported out of the cavity again via the gas pressure before solidification (Figure 2.37). The melt can either be pressed back toward the machine nozzle (melt back-flow process) or into an additional space next to the actual cavity (secondary cavity process). Particularly for components that are not only tubular but also have thin-walled areas, a conventional holding pressure can be applied before the blow-out process.

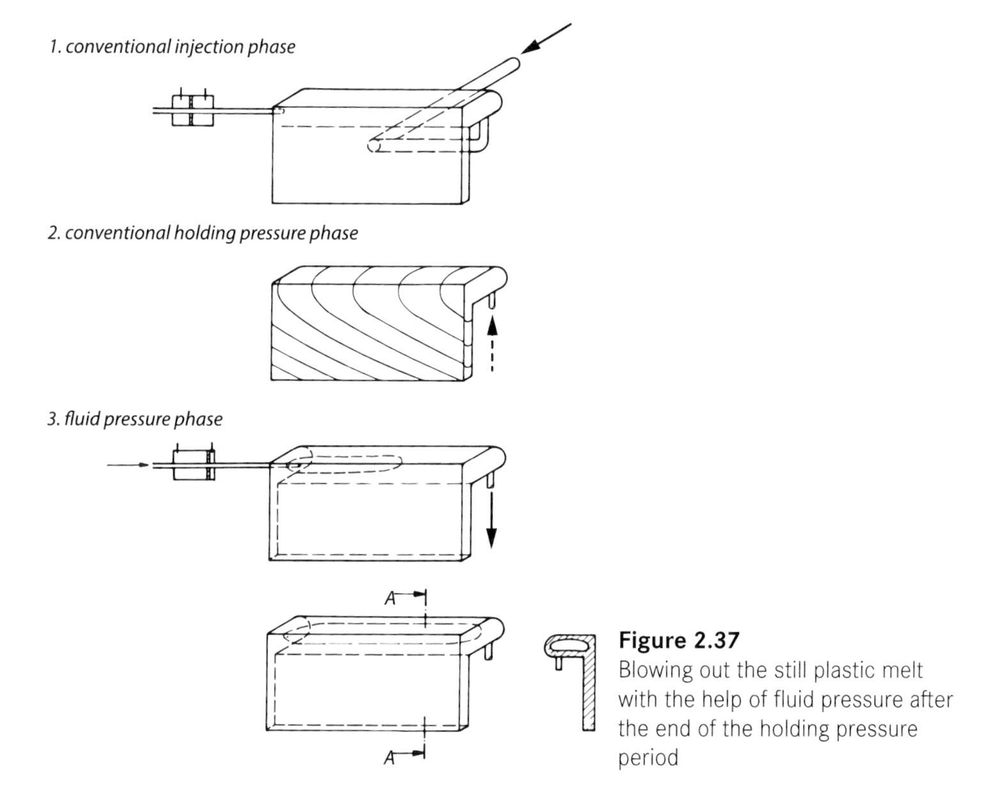

Figure 2.37 Blowing out the still plastic melt with the help of fluid pressure after the end of the holding pressure period

2.5.2.2 Component Design

Aspects of component design for fluid injection technology processes

The following questions are of importance for the designer:

- How are the fluid channels to be designed?
- For the cost estimation:
  - How heavy does a FIT component become?
  - What cycle times can be expected?

Include fluid channels in the design as flow aids

The designer must consider that a fluid channel provides flow assistance during the manufacturing process. The negative example of a ribbed plate (Figure 2.38) shows the course of filling from the center of the plate over time. When the melt is injected in the thin area of the plate and reaches the thicker ribs, it flows faster in the rib, because of the larger cross-section here. As it progresses, leading melt in the rib will now flow into the thinner area and join the melt that is flowing only in the thinner area. Towards the end of the filling process, even some flat areas will be enclosed by melt, resulting in air pockets and possibly a diesel effect with burns.

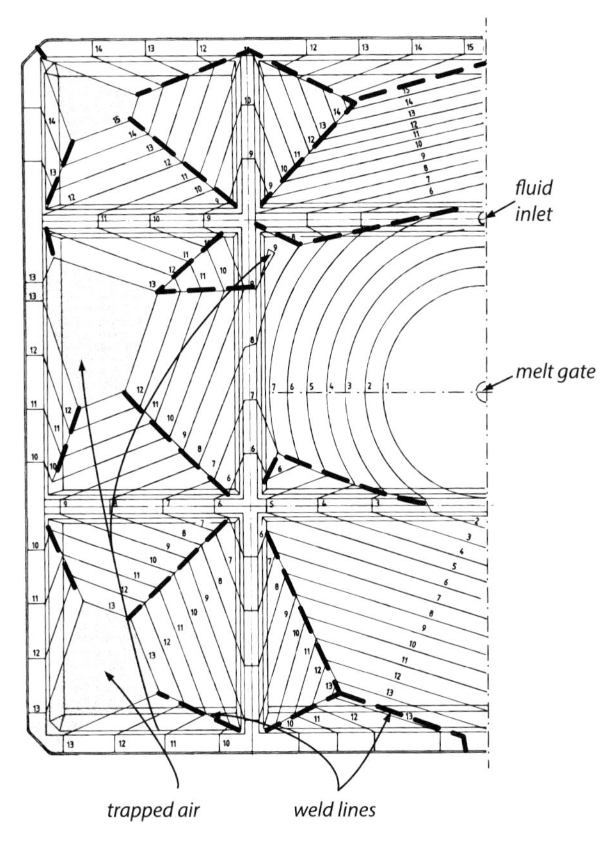

Figure 2.38 Filling behavior of a ribbed plate with air pockets and weld lines

Designing for FIT-2

Example of advantages to be considered when developing a component for the FIT process

The development steps for FIT-2 are illustrated using the example of a plate ribbed on three sides (Figure 2.39):

Ribs as flow assistance

- Consideration of ribs as providing flow assistance. The areas to be filled last should, if possible, border the parting line so as to avoid the diesel effect. In the case of thin components with fluid ribs, it should be remembered that filling in the first step is carried out exclusively with melt. In this case, a simplified filling simulation conducted with a commercial standard simulation program may prove helpful.

Fluid provides residual filling – an estimate largely facilitates localization of hollow spaces

- In the second step, the fluid flows in and provides the residual filling. The fluid will preferentially flow through the thicker ribs because the pressure demand is lower here. As an estimate, it can be assumed that after fluid injection the fin cross-section is only 60%. For a component with a rib fraction of, for example, 20% of the total area, the extent of partial filling is correspondingly about 80% (thin-walled area), plus 0.6 times 20% (rib fraction), i.e. about 92%.

- Choosing the location for fluid injection is a challenge. If the fluid is close to the melt gate, the intended fluid channels must not be filled with fluid to the end, because the fluid always takes the path of least resistance. If necessary, the fluid will leave the channel at the end of filling and flow toward the end of the flow path (the thin area in Figure 2.39). Alternative sprue and injection locations (Figure 2.40) change the scope for complete cavity formation. In the event of an unfavorable choice, massive thick spots and thus long cycle times and sink marks must be expected, because then the intended ribs cannot become hollow at all.

Challenge in the design process: "fluid always takes the path of least resistance"

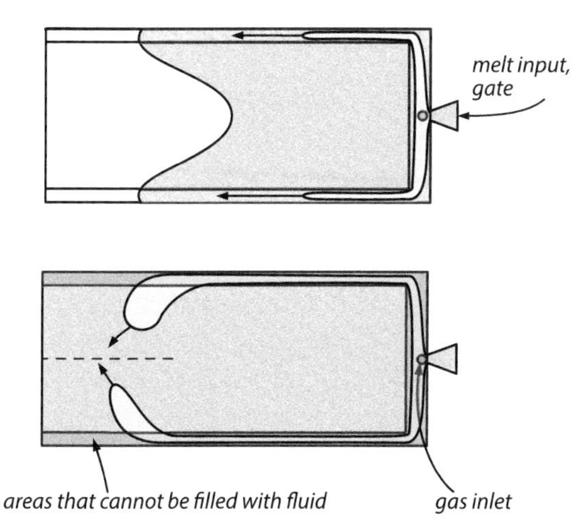

Figure 2.39
Unfavorable choice of fluid injection in the FIT-2 process

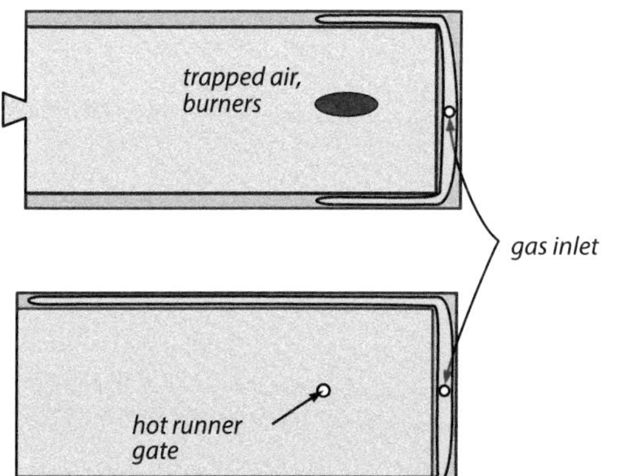

Figure 2.40
Alternative sprue and injection locations (top unfavorable, bottom good)

- Because the fluid applies the holding pressure, consideration must be given to the shrinkage potential in the thin-walled area. In concrete terms, this means that, if the fluid channels are very far apart, there is a great risk of the finger effect (Figure 2.41). The gas presses on the shrinking plastic melt in the thinner areas and pushes itself forward into these areas while the plastic is still warm enough.

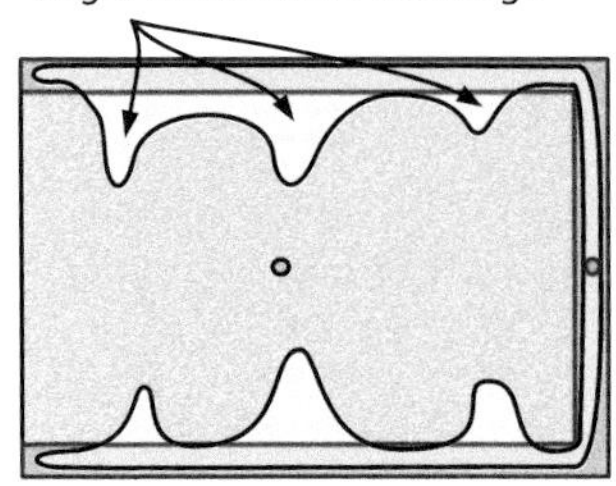

Figure 2.41
Finger effect due to unaccounted-for shrinkage potential

- When there are many intersecting ribs, it must be remembered that fluid channels will never meet and join. At the point of meeting, i.e. in the area of weld lines, melt basically remains in the rib as a solid seal.

Designing for FIT-3

FIT-3 is less complicated

Development is simpler for FIT-3. If the melt is blown back in the direction of the screw cavity (melt backpressure process), the gate must be selected large enough so that blowout is not rendered impossible by premature freezing. If several cavities are to be blown out (Figure 2.42), a time sequence should be planned. With two or more cavities, it is unlikely that the fluid channels will form at the same rate. Thus, one fluid channel may reach the central sprue sooner, preventing the other cavity from being blown hollow. Timing control can be achieved with a controllable shutter slide. Initially, one of the two cavities is blown hollow when the slide gate is open. The progress of cavity formation can be monitored via the return signal from the screw. After actuation of the slide gate, the pressure for the next cavity can be initiated while the pressure in the first cavity is maintained. In the case of a handle, the blow-out process takes approx. 14 s, because the fluid pressure, at approx. 200 bar, is lower overall than the injection pressure, and thus pushing back of the melt is slow. During this time, the solidified boundary layer grows in the neighboring cavity that has not yet been blown out, so that the parts to be subsequently blown out tend to become somewhat heavier.

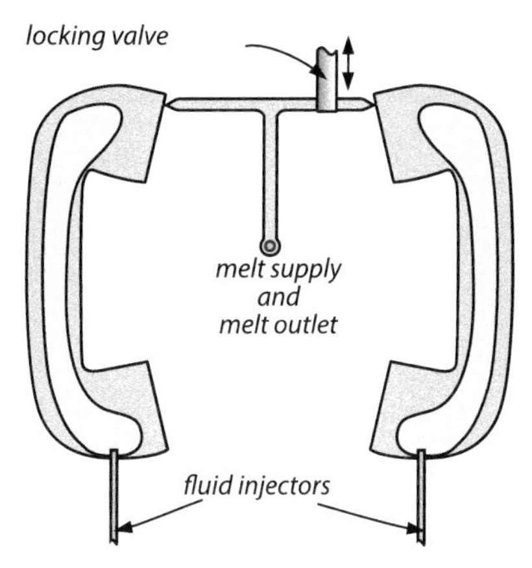

Figure 2.42
Arrangement of two cavities in the melt back-pressure process

Realization via a pin at the end of the fluid channel

If the secondary cavity process is selected, it must be ensured that the secondary cavity is not already filled during the holding pressure phase. This is feasible with a tunnel connection to the secondary cavity via a sealing pin (Figure 2.43). At the end of the fluid channel is provided an approx. 2 mm thick pin, actuated by a core back, which projects into the cavity. Because of its low mass and the surrounding melt, the pin is very hot, so that hardly any solidified surface layer forms here. When the pin is withdrawn at the start of the fluid pressure phase, the surface of the component is torn open at this point and the melt can flow freely into the secondary cavity.

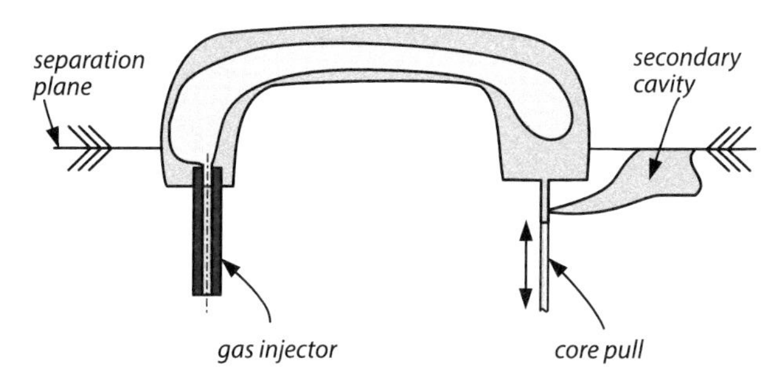

Figure 2.43 Design of a closable secondary cavity for FIT-3

Independent shut-off cavities

Complicated arrangements of cavities are possible via independently actuated, lockable cavities. The part in Figure 2.44 has four connected cavities. At the design stage, it can already be planned that, for example, channel L1 is the first to be blown out. If this is the case, the shut-off gates for the other branching channels remain closed. The other gates are opened either time-controlled or via a pressure signal at the end of the secondary cavity V1.

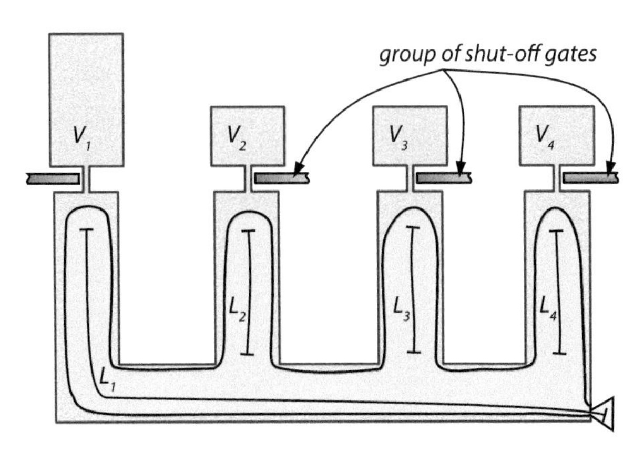

Figure 2.44 Design of a complex fluid channel structure in the side cavity process

Aspects that should be considered for secondary cavities

The following must be observed for secondary cavities:

- Their design should not be too thick, because they fill with a delay after the injection process and the melt reaching here must be cooled down far enough before demolding.
- The volume of a secondary cavity should be adjustable from the side of the opened cavity. The volume V of the secondary cavity corresponds to the fluid channel length L times the cross-sectional area of the channel. The cross-sectional area is the result of the process and is due to flow conditions that cannot be accurately predicted in the design phase. Adjustability of the cavity therefore allows the fluid channel length to be optimized during process start-up.

3 Molds

Simple molds are more reliable in production

It sounds banal: in order for an injection molded part to be produced, an injection mold is essential. The designer of a plastic component must pay particular attention to this, because a complex component inevitably requires a complex injection mold. The higher his/her level of knowledge of mold technology, the more likely it is that the designer will be able to ensure the simplest possible molds. After all, in the end it is a matter of reproducing a manufacturing process with high volumes, and here too the principle applies:

> *"Perfection is not achieved when nothing more can be added, but when nothing more can be left out." – Antoine de Saint-Exupéry*

The more individual parts a machine has, the higher the risk of failure and thus the product costs increase.

Illustrations of the molds are rotated through 90° for the mold maker

For the design of an injection molded part, it is expedient if the mold manufacturer or mold maker is contacted at an early stage. It is easier to deal with each other when the designer presents his/her ideas in a way that suits the mold maker. For the production team, the mold is always shown in its installed orientation, i. e. with the parting line vertical in the picture. By contrast, for the mold maker, the mold is always rotated through 90°. Each mold half consists of various individual parts, which are easier to assemble in this position. For this reason, in this book all figures and sketches of molds are oriented accordingly.

This also explains the term mold installation height, which is the distance between the outer dimensions of the nozzle side and the ejector side. In mold dimensions, this dimension is also called the mold height. With regard to a mold installed in a machine rotated by 90°, there is then the vertical dimension length and the horizontal dimension width.

3.1 General Tasks and Functions

Standard mold, also called two-plate mold

Figure 3.1 shows a standard two-plate mold. By plate is usually meant a plate package. Here, a package consists of one clamping plate each for fixing in the clamping system of the injection molding machine and the cavity plates. In a two-plate mold, the separation of the two halves forms the parting plane between the "A side" (alternative names: moving side, core half, ejector side) and the "B side" (fixed side, cavity side, nozzle side).

The component itself is reproduced in the cavity. The melt is usually introduced centrally through the plates of the nozzle side. The heat from the melt is dissipated via the cooling liquid, which flows through the various cooling channels. After the mold opens, the largely cooled component can be demolded with the aid of an ejector mechanism.

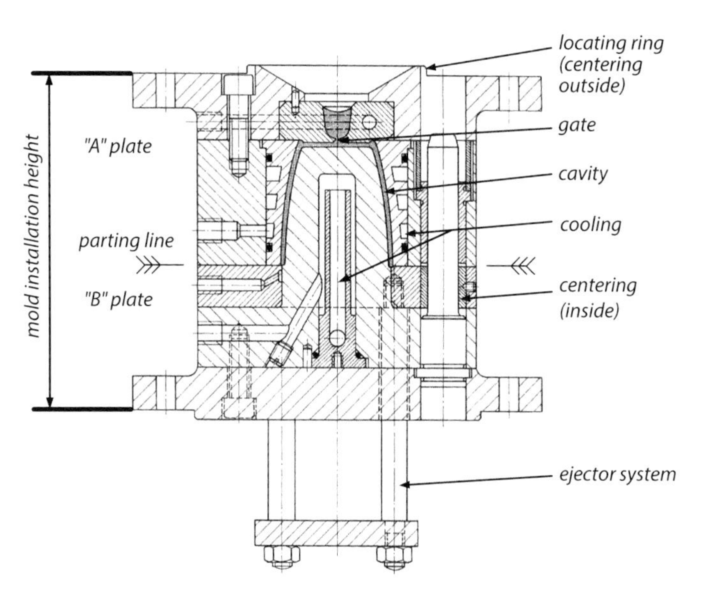

Figure 3.1 Cross-section of a simple mug mold, showing essential functions [image source: Mennig, "Werkzeuge für die Kunststoffverarbeitung"]

Inserts in cavity plates

The cavity either is machined directly into the mold plate or a suitable cavity insert is used (Figure 3.2). Reasons for using the insert variant are as follows:

- The parting line is specified via the contour of the component and is not necessarily flat. With a curved parting line, the machining effort is greater if the cavity is incorporated directly into the mold plate.

- If the components have a large height, very thick mold plates would be necessary, from which a lot of material would have to be removed. Thinner mold plates can be used with inserts.
- If any changes are made to the cavity, replacing an insert is always cheaper and faster.

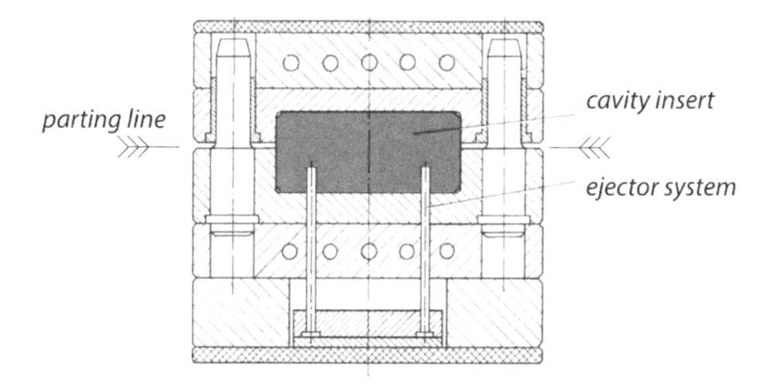

Figure 3.2 Master mold with exchangeable cavity insert. Here, only the installation space is shown – the insert itself is usually in two parts (nozzle side, ejector side) [image source: Mennig, "Werkzeuge für die Kunststoffverarbeitung"]

In addition to the above-mentioned advantages of cavity inserts, the following must be taken into account:

- The cooling should also routed through the inserts. In this case, the cooling channels are best routed into the insert from the back side, i. e. from the rear of the insert in the direction of the clamping force, so that the cooling channels can be sealed most easily by means of O-rings and the inserts can be mounted easily.
- The machining effort for the insert and the recess in the cavity plates is greater than when the cavity contour is machined directly into the cavity plate. This increases the costs. A cavity insert is not useful in every case.

Normally, the parting plane is largely perpendicular to the opening movement. In the case of shear edge molds, the mold halves in the area of the cavity separate exactly parallel with the opening direction (Figure 3.3). This makes it possible to variably change the cavity height. This type of mold is used to reduce the cavity volume by means of a closing movement during melt cooling. As a result, the cooling plastic can be kept under pressure over a wide area; this is a injection compression process and is useful for predominantly flat components.

Shear edges are surfaces which part parallel with the opening movement

Polymer melt can flow through gaps > 0.03 mm at high pressure. Shear edges should therefore be designed with replaceable shear edge strips wherever possible, because they must have very few gaps overall. Such strips are therefore at risk of wear and can be replaced or serviced with minimal effort.

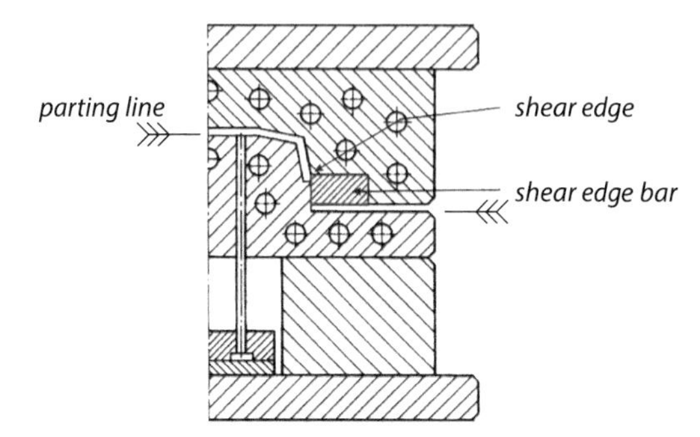

Figure 3.3
Shear edge mold with replaceable shear edge bar for variable cavity height [image source: Mennig, "Werkzeuge für die Kunststoffverarbeitung"]

3.2 Manufacture and Costs

Important questions governing the choice of mold concept

For development of the mold concept, the following aspects need to be addressed:

1. *Component orientation/separation:* The component should be placed in the mold such that easy demolding is possible. In the best case, a two-plate mold can then be used without additional demolding direction via transversely movable elements.
2. *Mold design* (single face mold, mold with sliders, ...): Where the component is unfavorably positioned and also in the event of an unskilled design, simple demolding is not possible. Usable mold concepts for complex demolding are described in Section 3.6.
3. *Number of cavities:* The number of cavities n determines the mold and machine size. Machine size in this context always refers to the necessary clamping force. Injection molding machines also become larger overall with increase in clamping force, and larger molds can then be installed. Several cavities in the parting line may make the mold wider and longer. It may then be necessary to select a machine that offers a larger installation space. Usually, the necessary clamping force determines the machine size. The possible number of cavities or the necessary clamping force can be simply estimated as follows:

$$n_{cavity} < \frac{F_{clamp}}{\bar{p}_i A_{proj}} \quad \text{(assumption: } \bar{p}_i < 300\,\text{bar)} \tag{3.1}$$

   where A_{proj} is the projected area of the components. By this is meant the shadow cast in the direction of the parting plane.
4. *Alignment of the cavities:* The layout of several cavities in the parting line determines the gating system and the mold size (see Section 3.4.1.1).
5. *The design of the cooling system and ejector system:* Both systems require space in the mold and must not collide.

6. Mechanical processing accounts for approx. ¾ of the total manufacturing costs of a mold (Figure 3.4). To a large extent, the mold consists of standardized purchased parts that largely do not have to be machined at all. Therefore, these costs are mainly incurred for the production of the cavities. Design costs here relate less to the cost of part development and more to the cost of making changes to the part design and construction of the mold. Changes to the component shape often become necessary when the article designers have not given enough consideration to possible implementation in a mold, e.g. the article has insufficient draft angles.

The manufacturing costs of a mold are largely determined by the production of the cavities

7. Material costs account for approximately 10%. These costs refer to the standard purchased parts and the material for the cavity plates. Due to misplaced thriftiness, cheaper but less suitable materials are often purchased for this purpose (see Section 3.2.3).

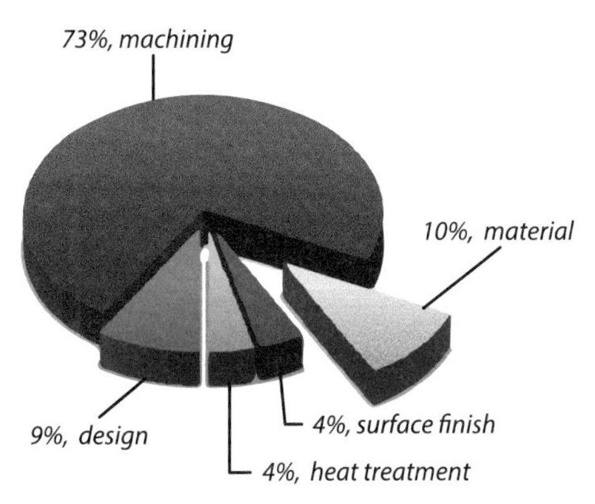

Figure 3.4
Breakdown of the costs of an injection mold [image source: "Handbuch der Kunststoffstähle", Edelstahlwerke Buderus AG]

The designer should be aware that the cost of a mold is largely determined by the number of cavities. A 2-cavity mold is about twice as expensive as a 1-cavity mold, even if the cavity dimensions are small enough that the outer dimensions of the mold remain the same. The cost of building a mold is dominated by the cost of machining, and most machining directly concerns the cavity. The mold elements further away from the cavity (mold plates, centering elements, ejector base plates, …) can be purchased as standard parts ready for installation. The costs of the mold steel only come to approx. 10%.

3.2.1 General Machining

Cost predictions are necessarily very rough. Very difficult to estimate are the costs for the various correction cycles after initial testing (Figure 3.5). The more complex the plastic component, the greater is the risk of changes. These changes par-

A mold is often revised several times before it is released

ticularly affect the component dimensions, because shrinkage and process-related component warpage are very difficult to predict accurately.

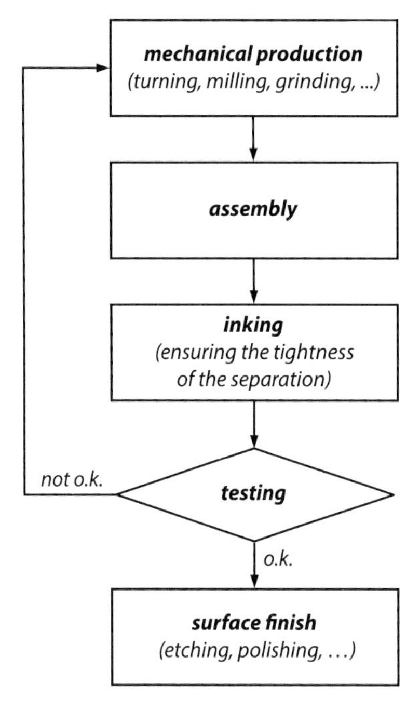

Figure 3.5
Steps involved in mold production

Strategy for low reworking costs

It should be remembered that mechanical reworking of the cavity is equivalent to increasing the size of the cavity (Figure 3.6). There are two possible approaches:

a) The inexpensive type of rework is machining, i. e. further removal of mold steel. The shape of the component created in the initial testing is assumed to be smaller and can be adjusted or enlarged to the correct dimension in the correction phase.

b) Molds are increasingly being designed with multiple inserts so that only partial areas of the cavity have to be reworked or recreated in the event of correction.

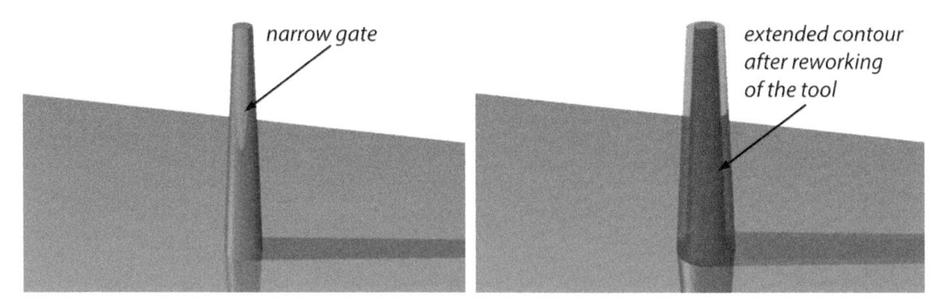

Figure 3.6 Mechanical reworking enlarges the cavity [image source: Protomold]

In the mold manufacturing workflow, the spotting step involves sealing the parting surfaces, especially the parting line. The seal must be so good that during injection the plastic melt cannot flow into the parting line and form flash, while at the same time the allowing air in the cavity to escape. To this end, the fully assembled mold halves are rubbed against each other in vertical clamping units (Figure 3.7). A spotting color previously applied to one side will transfer to tightly contacting mating surfaces. Any mating area that does not become colored therefore does not lie flat. After appropriate reworking and repeated checking with spotting color, the parting line should be completely flat. Gap widths of 0.03 mm are tolerable here, because the plastic melt cannot flow through these gaps under pressure and may form flash in the area of the mold release.

Spotting for ensuring accuracy of fit

Figure 3.7
Spotting press with clamping table that can be tilted for fitting work [image source: Millutensil]

3.2.2 Surfaces

Surface finishing is the last step of mold production and is done when all the dimensions are in order after testing. The various finishes are:

Surface finishing is the last step

- *Unprocessed* – Largely the non-visible surfaces of the back of the component, machining marks from, e. g. milling, are visible on the component.
- *Polished* – Very high-gloss surfaces; this processing step is very expensive because a high level of manual effort is required.
- *Spark eroded (EDM)* – Somewhat textured surfaces with overall uniform roughness, giving a uniform gloss or matte finish.
- *Fine-textured* – Grained surfaces with, for example, a leather or wood look. There are different manufacturing processes for this texturing.

3.2.2.1 EDM – Electrical Discharge Machining

EDM for fine contours and defined surfaces

The EDM process is based on the principle of spark erosion. Depending on the amount of material removed, EDM can be used either to change only the surface texture of the cavity or to produce larger contour areas. The EDM process is more suitable than milling, especially for free-form surfaces, narrow recesses, and recurring patterns.

An electrode is placed in a dielectric near the surface of the workpiece (Figure 3.8). A short electric arc is generated via an electrical voltage potential, the energy of which melts the surface of the workpiece and blasts away small amounts of material. The voltage potential is generated in short pulses so that small pits are created on the surface (Figure 3.9).

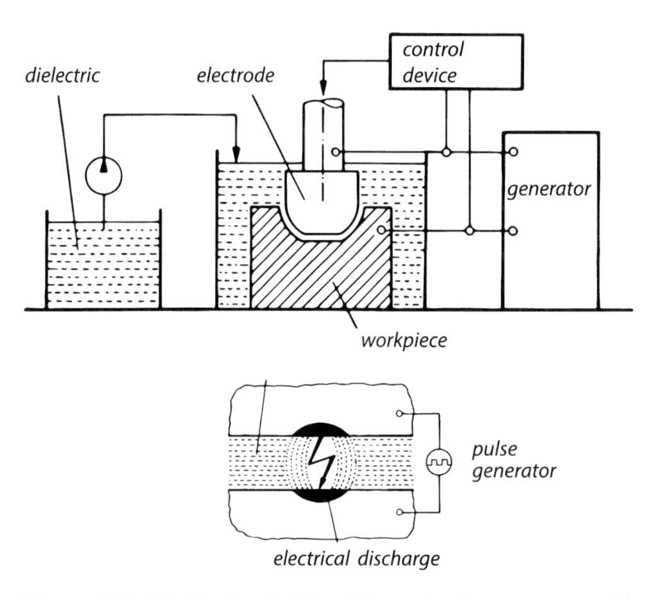

Figure 3.8 Principle behind die-sink erosion [image source: Menges]

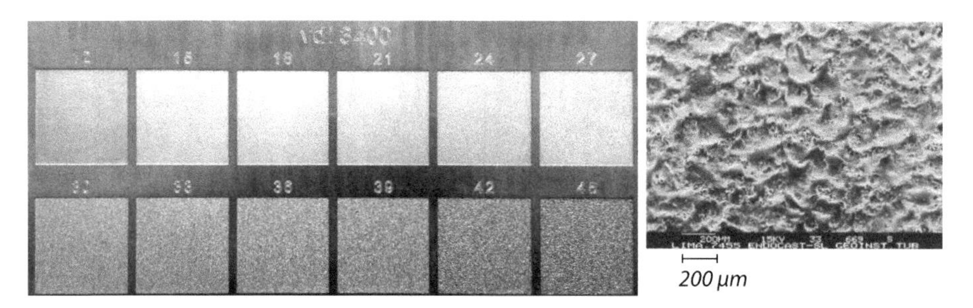

Figure 3.9 Texture of an eroded surface with roughness according to VDI 3400; SEM close-up of texture No. 33

A distinction is made between wire-cut and die-sink EDM. In wire-cut EDM, a thin wire is fed as an electrode through a steel plate and thus cuts out very fine vertical and largely sharp-edged contours.

A prerequisite for die-sink EDM is an electrode which has the shape of the negative image of the surface and is either made of copper or graphite and has been produced in advance, e.g. in a milling process. These electrodes can create deep indentations in steel plates, and the contour of the created surfaces can be chosen very freely. Electrodes can also be used to produce surfaces with a roughness defined in accordance with the recommendation of VDI 3400.

EDM textures can create defined roughness

Several surfaces can be machined with one electrode. Thus, a larger surface that is to be given a special surface texture can be produced in several operations with a smaller and hence less expensive electrode. The actual eroding process is well automated and can therefore run unattended.

3.2.2.2 Etching

In contact with a pickling agent (usually acid), a metal surface can be chemically attacked, whereby individual atoms are dissolved out of the metal surface. Metal erosion can be specifically prevented by covering part of the surface with wax, for example.

Photochemical etching (Figure 3.10) uses a negative mask of the desired texture. UV light is used to expose a coating on the metal surface and this negative pattern is transferred to the surface. The coating can be washed off the areas not covered by the mask and not exposed. During the subsequent etching process, depressions are created and the positive of the texture is formed on the surface. Even though the process sounds simple, it is elaborate and takes a lot of time.

Working steps for etching processes

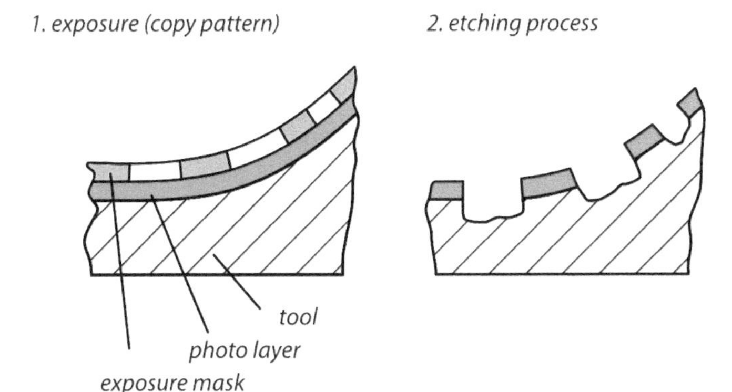

Figure 3.10 Photochemical etching for surface texturing

The textures can be highly diverse (Figure 3.11). Because in many cases an imaging process is used upstream, it is also possible to produce very large-area components with a leather look, even if correspondingly large original leather samples are not available.

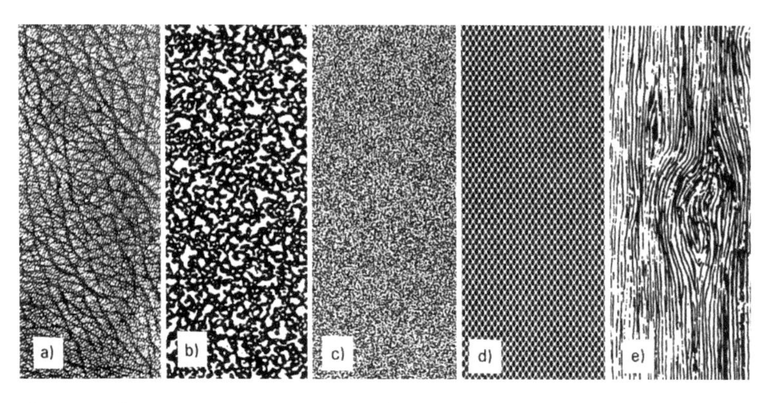

Figure 3.11 Typical etching textures [image source: Mennig]
a) Leather grain
b) Imitation hammer blow
c) Technical fine texture
d) Lattice texture
e) Wood texture

3.2.2.3 Laser Texturing

Surface ablation with energy of an aligned laser beam

Texture can be created directly with a controlled laser beam that scans the surface. The energy of the laser beam vaporizes a small part of the surface at the point of impact, creating small depressions in a manner similar to spark erosion. Because the laser beam can have a width of down to 40 µm, very fine textures are possible (Figure 3.12). The textures that the laser ablates are compiled in advance on the computer in a similar way to photochemical etching and are transferred to the surface of the component.

The advantages of this technique over etching are as follows:

- Very fine and sharp-edged textures can be created.
- The textures can have a very different height profile.
- They can be created on almost any three-dimensional surface.

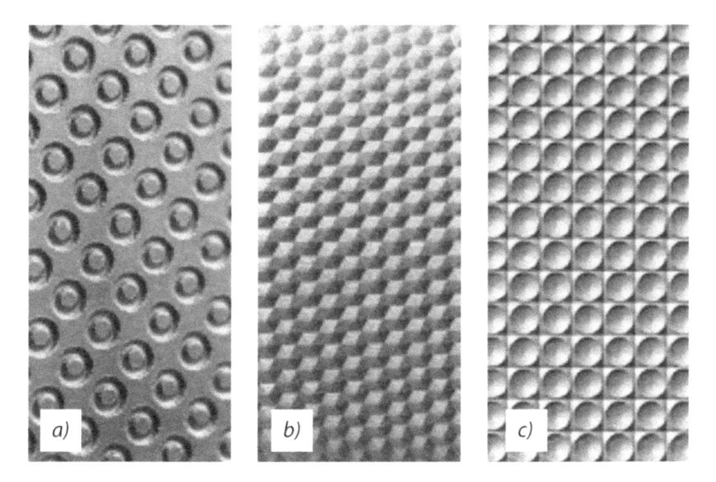

Figure 3.12 Examples of laser textured surfaces [image source: DMG Mori]
a) Circular grooves
b) Honeycombs
c) Knobs

3.2.2.4 Ceramic Surfaces

Impression of natural surfaces with ceramics

Ceramic coatings can be used to create similarly flexible textures with a highly variable height profile. With this process, it is possible to transfer surfaces directly onto a plastic component. To this end, the original surface is first molded with silicone, which creates an exact negative texture (Figure 3.13). This is subsequently molded again in silicone, creating a duplicate of the original, which can be represented as the texture of a flat surface due to the pliability and ductility of the silicone. From this duplicate, a negative replica is created using a ceramic paste. This ceramic layer too is initially still flexible, so it can finally be transferred to non-planar mold surfaces, where it is baked at temperatures of approx. 250 °C.

The ceramic layer has good wear resistance and can be used to make a high number of impressions in plastic melt without losing texture. In addition, it acts as a heat brake due to its lower thermal conductivity. From layer thicknesses of approx. 0.5 mm, solidification of the melt on first contact with the mold is noticeably slowed down. As a result, less injection pressure is required to fill the cavity. In addition, the melt remains soft somewhat longer and is better able to reproduce very fine textures. However, the cooling time does not slow down measurably, because the layers are altogether too thin for thermal insulation.

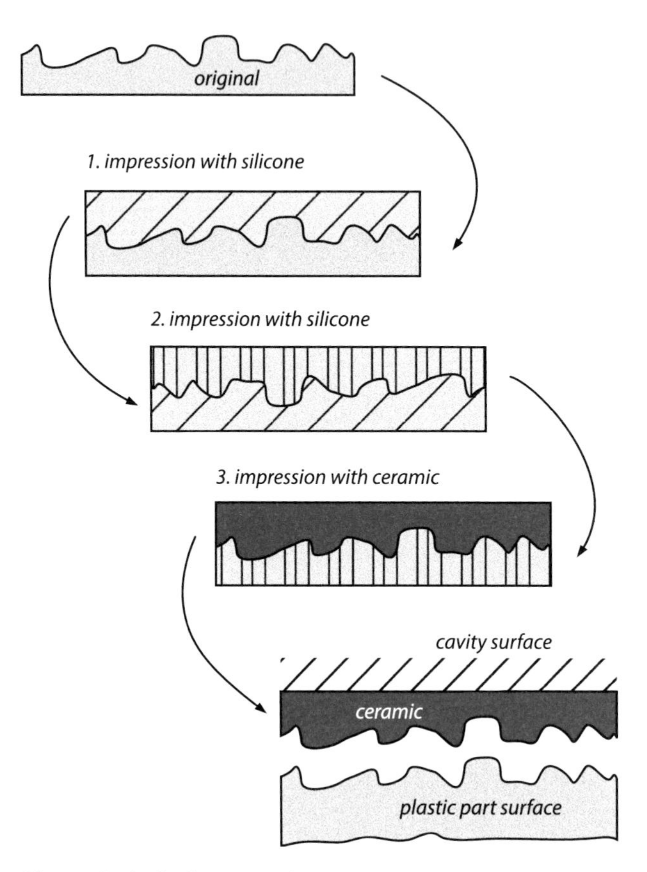

Figure 3.13 Surface texturing over ceramic surfaces

3.2.3 Steels

Requirements on mold steel

With regard to the costs of a mold, the roughly 10% share accounted for by materials is quite low. Nevertheless, the required properties of the mold steel are often underestimated, with the result that unsuitable and inexpensive grades are selected, which can even increase the mold costs in the long run. The overview of important properties given in Figure 3.14 is divided into machining and usage. First of all, good machinability is important for the manufacture of the mold. Therefore, aluminum is often used for test molds, because high cutting speeds are possible here during milling. After successful initial testing, the cavity inserts are given the required surface finish, often by etching or polishing.

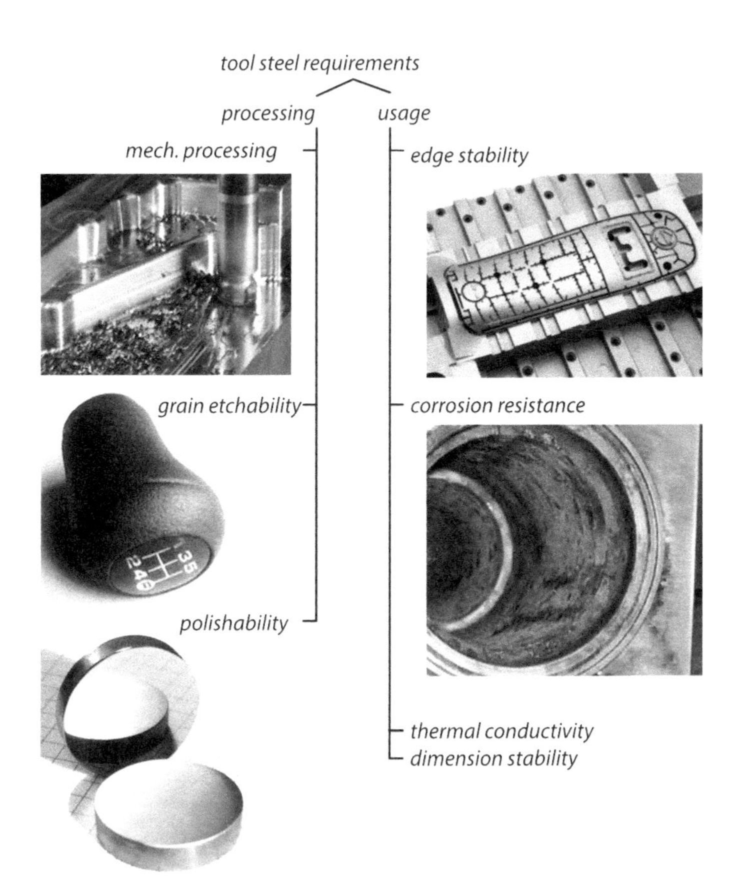

Figure 3.14 Requirements on mold steels [image source: Mayerhofer, Böhler GmbH]

The polishing process can be noticeably prolonged by material inclusions. In many cases, inclusions are very stable oxides (compounds of alloying elements with oxygen) which form as the steel is being produced. During polishing, these particles resist the polishing agent, resulting in visible elevations in the polishing direction (Figure 3.15). Especially with regard to good polishability, there are powder metallurgical steels available which are produced not from the melt, but via a sintering process of very fine powders. This reduces the formation of inclusions or uneven distribution of alloying elements.

Homogeneous steels can be polished better

The steels used for injection molds are divided into cold-work and hot-work steels (Figure 3.16). Cold-work steels can be hardened to increase wear protection and may then only be exposed to moderate temperatures below the tempering temperature. Higher temperatures cause tempering of the steel, which reduces hardness and increases toughness. Hot-work steels attain their maximum hardness only after a defined tempering process. They are therefore suitable for temperatures below the tempering temperature and are thus initially suitable for use in the cavity area.

Hot-coldwork mold steels for higher temperature requirements

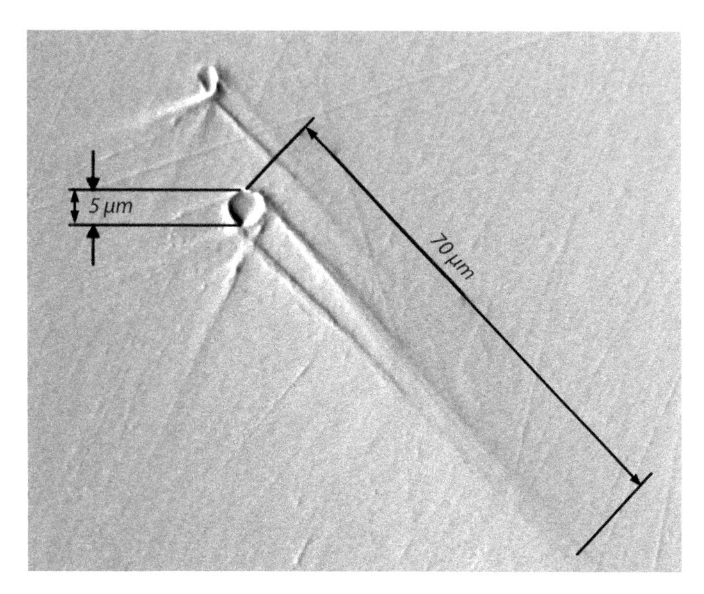

Figure 3.15 "Comet" polishing defect caused by deposits in the mold steel [image source: Mayerhofer, Böhler GmbH]

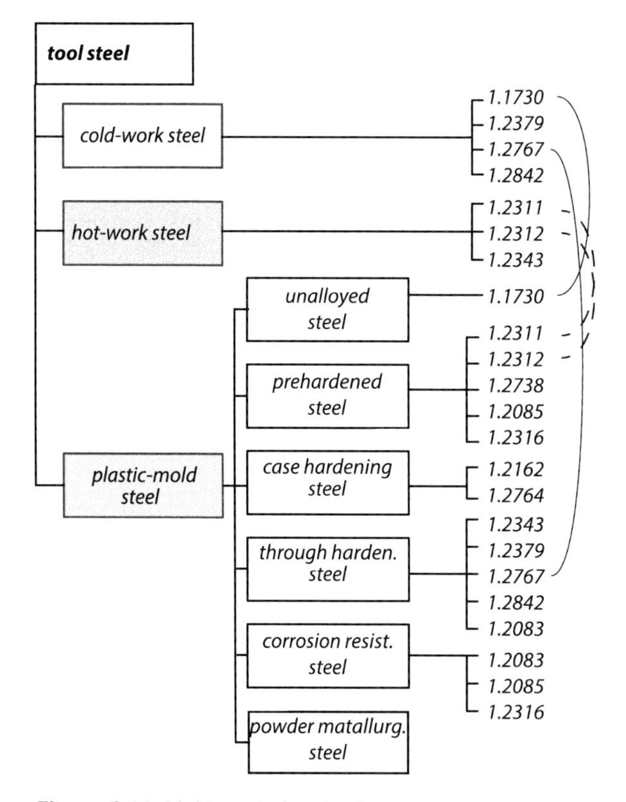

Figure 3.16 Mold steels for plastic molds

The requirements imposed on the various components of the injection molds (Figure 3.17) call for different steel grades, depending on the mold area:

Special requirements, depending on the mold area

- Unalloyed steel is inexpensive and suitable for mold areas without special requirements; these include the parallel and the ejector plate.
- Prehardened steel in the as-delivered condition already has a strength (yield, yield strength) of over 1000 N/mm^2. The prehardened steels are supplied with their properties and are not intended for further heat treatment. They are used for mold areas subjected to higher loads: the clamping plate of the "A-side" and the "B-side", the ejector base plate and the support plate. Under the effect of the injection pressure, there is a high mechanical load and thus possibly deformation, especially in the area of the travel of the ejector plate and the lack of support of the central mold area in the area of the boring in the nozzle plate of the machine for the machine nozzle. As no further heat treatment is provided, these steels are also suitable for dimensionally accurate mold areas, because distortion is not to be expected due to the absence of subsequent hardening.
- Case-hardened steel is low in carbon and easy to process in the delivery condition. It is only through extensive heat treatment, which involves carburizing and subsequent hardening, that a high surface hardness is obtained with a soft or tough core. These steels are used for cavity cores and for areas with small surface engravings. However, the possible distortion of the cores during heat treatment is problematic and requires reworking.
- Through-hardened steel has great hardness, even at large wall thicknesses, thanks to simple heat treatment combining austenitizing and subsequent hardening. Post-processing with erosion processes is not critical, because no consideration needs to be given to the thickness of the hardening layer. These steel grades are suitable for applications with a high risk of wear, e.g. plastics with a high glass fiber content.
- Corrosion-resistant steel is used when aggressive plastics such as PVC or flame-retardant materials are to be processed. In such cases, the non-cavity areas are often also given a corrosion-resistant finish.

ESR quality and PM steels

ESR (Electro-Slag Remelting) grades or PM steels are particularly suitable for applications requiring a polished mold surface. This production method affords particularly pure steel grades with low levels of impurities. The impurities complicate the polishing process because they are responsible for the occurrence of grinding defects (Figure 3.15), which make repeated polishing processes necessary. PM stands for powder metallurgy, a steel production process in which pressed metal powder is sintered at high temperatures. Due to the absence of the melt phase, no melt precipitation takes place in this manufacturing process, resulting in the formation of very uniform microstructures that are easy to polish.

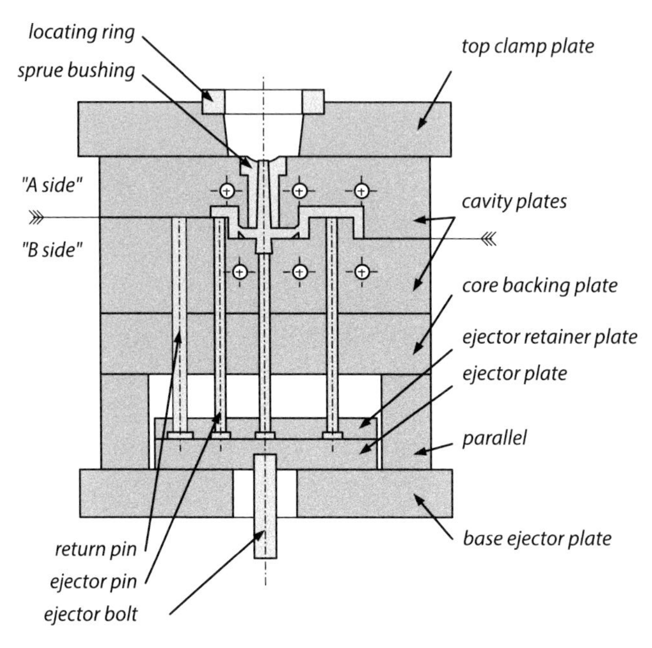

Figure 3.17 Identification of the various mold plates

■ 3.3 Standard Elements

Standard parts for standard tasks

Molds consist to a large extent of standardized components – only the area in direct contact with the melt requires independent solutions. The circled area (Figure 3.18), which is not covered by standardized components, shows the:

- *Cavity:* which is incorporated into the cavity plate or an insert;
- *Temperature control:* which usually consists of various deep drill holes;
- *Ejector:* for which holes must be drilled in the cavity plate.

Reasons for the use of standard parts

The advantage of these standard parts is the standardization of dimensions, types, and procedures. The goals are as follows:

- Standard components simplify spare parts inventory. Some elements are subject to a certain amount of wear, e.g. the sprue bushing is mechanically susceptible and must be replaced occasionally. This also includes the ejector pins.
- Standard parts suppliers mass-produce standard elements, so in-house production often results in higher costs.
- Consistent use of standard parts can reduce the variety of parts in a mold, which also makes assembly easier and saves costs.

- The design effort is reduced, because solutions are available for many known problems (see Section 3.3.1).
- In many cases, suppliers of standard parts offer special software packages for mold configuration in addition to the CAD data for the various components, so that the entire basic structure of the mold can be selected in a standardized manner. For simplification, parts lists and order lists are also generated at the same time.
- The focus in mold design can be placed on core competence. This includes, for example, the way difficult cavity areas are demolded or the development of a mold in anticipation of correction loops after initial testing. It also refers to the construction of cores from several individual parts, so that only parts of the mold need to be corrected or remade.

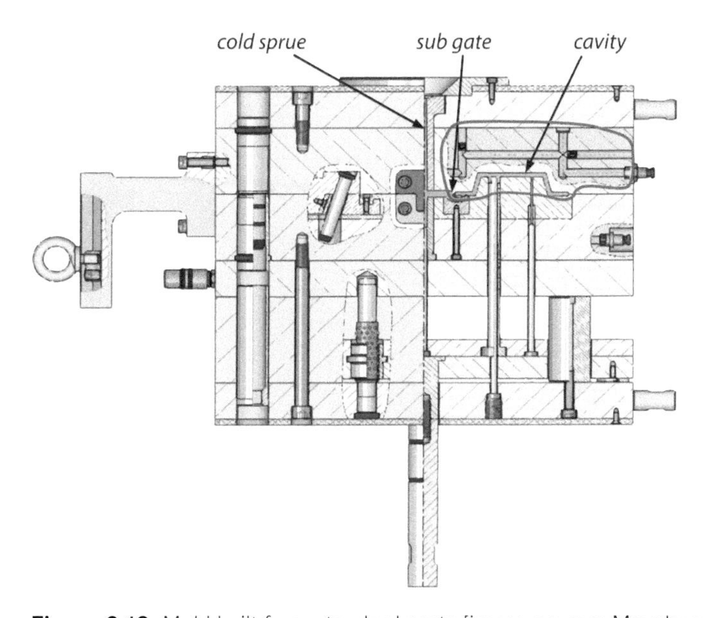

Figure 3.18 Mold built from standard parts [image source: Meusburger]

Standardized Purchased Parts

The range of available standardized purchased parts is highly diverse and can be roughly divided into:

- Plates
- Alignment elements
- Ejectors
- Cooling

Mold plates are available in graduated dimensions

Plates are available in different steel grades and in various sizes and thicknesses. In many cases, the plate sizes are selected in specified increments with regard to width and height, so that these plates are offered both undrilled and finish-machined. In the finish-machined version, the plates already have the holes usually required for the guide elements and the screw connections.

A mold consists of several plates placed one behind the other. Guide elements ensure correct positioning of the plates and also facilitate assembly of the mold (Figure 3.19). Guide bushes and centering sleeves are usually used on the ejector side, into which the guide pins on the nozzle side are inserted. The bushings and sleeves are each higher than the thickness of the plates, which aligns the plates with each other.

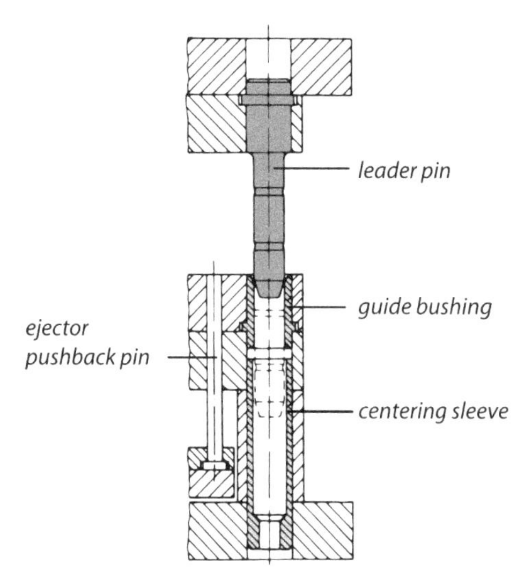

Figure 3.19
Ejector, guide, and centering element [image source: Menges]

Definition of guide and centering elements

A distinction is made between guiding and centering. Guiding provides the rough direction and is therefore uniaxial. Guiding elements ensure that the various plates of a mold are aligned and can be moved relative to each other without colliding. Centering establishes symmetry with respect to a specific point, e. g. a center point, for which two axes are considered. Exact positioning of the cavities relative to each other cannot be ensured without reliable centering.

Bushings and sleeves are used as guide elements. Bushings have close tolerances on the inside and outside. In this way, pre-centering of the cavity plates on the nozzle and ejector sides is possible when the mold is moved together. Sleeves have close tolerances only on the outside, so plates and wedges of a mold plate package can be aligned, i. e. separate ones for the "A-side" and the "B-side" respectively.

Centering for exact positioning of the cavity halves

The main task of guide elements is to guide. Centering has higher requirements overall and has tighter tolerances. Quite wide tolerances are sufficient for enabling the mold to be opened and closed without collision. To ensure that a component does not have offset edges in the parting line area, the cavity halves must be positioned very precisely relative to each other. The closer one gets to the cavity, the more important centering is. Areas further away can allow proper mold functionality even if they are offset from each other.

Thermal expansion in the mold is not uniform, and depends on the set temperature

It must be remembered that a mold may be warmer during operation than during assembly. Thermal expansion must not be hindered by the guide or centering. Slot interlocks positioned at the center of the mold that allow a change in the size of one half of the mold are useful here (Figure 3.20). It is imperative that round interlocks be correctly positioned and that they do not allow any size change due to different temperatures of the mold plates.

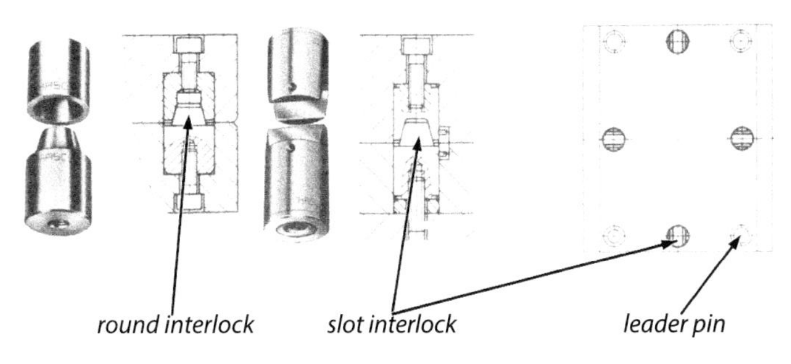

Figure 3.20 Centering of mold plates [image source: Menges-Mohren, “Spritzgießwerkzeuge”, Carl Hanser]

Centering of the cavity with wedges

Centering with flat wedges is appropriate for large molds and also for very tall and slender cores (Figure 3.21). Several heel blocks arranged at right angles (1-axis) produce biaxial centering. In the closing movement, this guide ensures final alignment of the core insert, which must be slightly floating. By adjusting the thickness of the wedge plate, the core can be moved relative to the opposite side so that it is precisely centered.

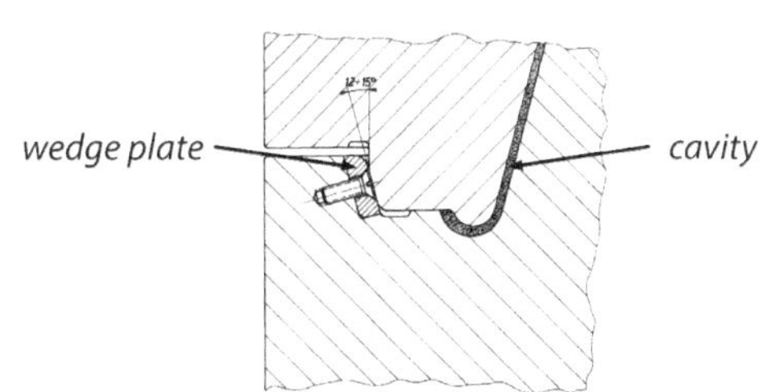

Figure 3.21
Centering of mold plates with taper lock [image source: Mörwald]

Demolding of undercuts with inclined sliders

In the simplest case, demolding is done with various ejector pins (see Section 3.6.1). For demolding undercuts, prefabricated inclined slider or inclined ejector units can also serve as a standard solution (Figure 3.22). The ejector moves at an angle via a link guide, thus making a lateral movement so that undercuts can also be demolded. Such tasks are very common and can be easily solved using standardized kits.

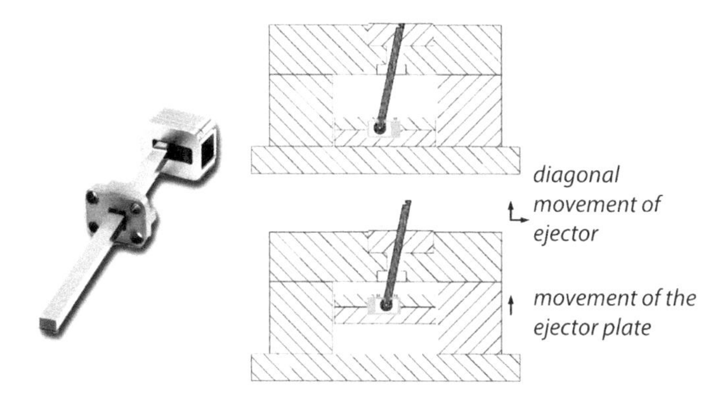

Figure 3.22 Inclined slider unit [image source: Meusburger]

■ 3.4 Melt Feed

Designations of melt-carrying areas

The melt enters the mold cavity from the machine's injection cylinder through feedthroughs, usually in the center of the nozzle side (Figure 3.23). In the simplest case, this is done unheated, so that the melt in the feedthroughs also cools and is demolded with the components. These systems are named as follows:

- *Sprue bar:* This is the direct connection between the machine nozzle and the component or a runner. It is conical for the sake of demoldability, and thickens toward the component. In the case of nozzle sides with large thickness, sprue bars can be considerably thicker at the connection to the runner than the components to be filled, which makes the cooling time very long.
- *Runner:* Where there are several cavities, at least one channel branches off at right angles from the sprue bar in the demolding plane.
- *Branch:* From the runner, channels branch off in the direction of the cavities that form the component. In principle, there is no need to distinguish between a runner and a branch.
- *Gate:* The connection between the branch and the cavity is often made somewhat smaller so that the runner system can be separated from the components more easily after demolding.

In the following text, there may be confusion with the term sprue because everything that is not a component is often referred to as a sprue. The sprue is necessary for production but is not used later and is either cut away or recycled. Frequently, therefore, the term sprue is used to refer to the entire system consisting of the sprue bar and runner.

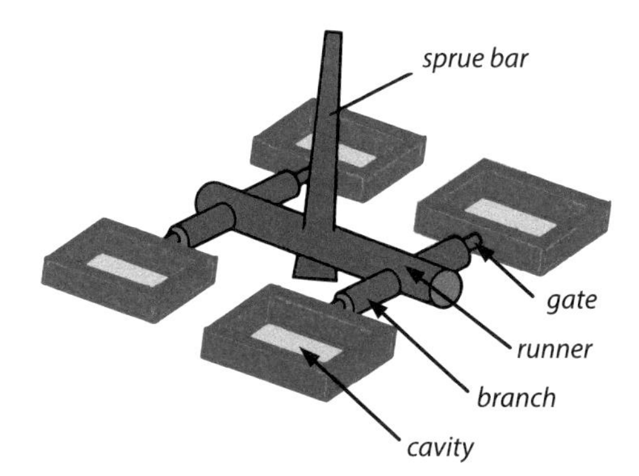

Figure 3.23
Melt feed via a cold runner system

Requirements imposed on the gating system

The entire gating system must meet different requirements:

- *Clean and easy separation* of the gating system from the molded part.
- *Demoldability:*

  Usually, the entire system is demolded when the mold is opened from the same parting line as the components. For demolding, the runner system must have cooled down sufficiently; otherwise it could crack, for example, at the transition of the sprue bar to the runner. The sprue bar would then remain in the mold and cause a process malfunction.
- *Holding pressure phase:* Sufficient time must be ensured.
- *Low pressure drop along the gating system:*

  Particularly if many cavities are to be reached via a runner system, the flow paths can become long. To ensure that sufficient pressure is still available for filling the cavities, the flow channels must not be designed too thinly.
- *Low heat loss:* Due to minimal surface.
- *Balancing of the melt flows:*

  In multi-cavity molds, the cavities should be filled at the same time. This applies either to the time at which the melt flows into the various cavities or, in the case of molds with cavities of different sizes, to the time at which the cavity is completely filled. Basically, this involves the pressure build-up which, at the end of the filling phase, ensures that the surface texture of the cavity is reproduced as faithfully as possible by the cooling plastic melt.

- *Small material ratio: Runner system to molded parts:*

  A gating system inevitably means a loss of material, even though thermoplastics can be remelted and the remaining plastic in the gating system can be fed back into a subsequent process. This generates costs and, in addition, the plastic is again subjected to thermal stress, so that the color of the plastic may change.
- *No damage to the melt:* Due to excessive shear or excessive residence time.

Partially contradictory design measures

Some of these demands require conflicting design measures (Table 3.1), and so compromise solutions must be found.

Table 3.1 Goals and Design Measures

Goal	Measure
Separation from component	Small gate
Long holding pressure	Large gate
Demoldability	Conical sprue bar, thickness similar to component thickness
Low pressure	Thickest-possible runner
Good balancing	Equal flow lengths to all cavities
Low mass	Shortest-possible flow lengths

A distinction is made in gating systems between cold runners and hot runners. In the simplest case, a cold runner can achieve a good compromise between the aforementioned diverging goals. In the hot runner variant, cooling of the melt in the runner is prevented by electrical heating. Apart from additional costs for heating elements and controllers, this also requires installation space in the mold. A large proportion of modern molds are now equipped with hot runners.

3.4.1 Cold Runners

Principle behind a cold runner

A cold runner is basically an extended cavity. It is therefore a melt-feed channel and also has to be demolded and cooled. Apart from the simplest form in which the sprue bar is gated directly onto a component, there is the possibility of gating several components via a runner system. Often this gate is thicker than the component and would therefore require a longer cooling time. However, if it is possible to reliably demold the sprue bar, the cooling time should, as far as possible, be based only on the quality of the component, for reasons of cycle time. The runner system can therefore still be quite warm and thus unstable at the time of demolding, as long as it can be demolded safely.

3.4.1.1 Cavity Layout

Star-shaped cold runner system

With regard to the layout and the requirement to produce as little sprue und runner volume as possible, a star-shaped runner system is an obvious layout (Figure 3.24). There are two decisive disadvantages here:

1. Due to the circular layout, the mold is initially round; this is not particularly useful for handling and storage, especially as molds are stored upright. Otherwise they have to be tilted, which is not easy with large molds.
2. In many cases, lateral movements of mold areas are necessary before demolding. A circular layout of identical components results in different directions of movement. This inevitably increases the number of drives and makes them more prone to malfunctions.

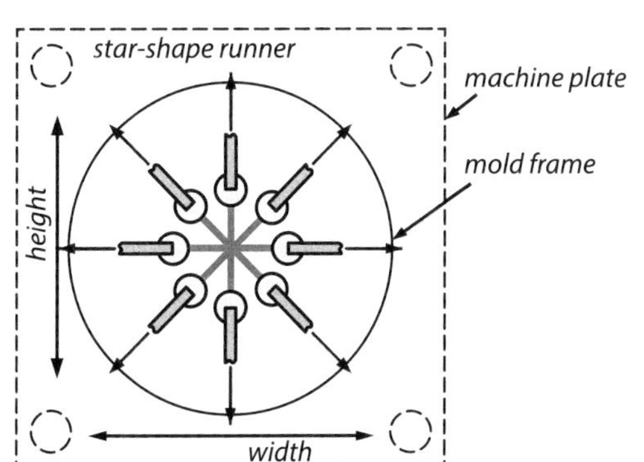

Figure 3.24
Star-shaped cold runner system

H-shaped cold runner system

Predominantly, several cavities are arranged in rows next to each other (Figure 3.25). Possible additional movements can be mechanically coupled via a common drive. In this way, the mold may become higher than it is wide, which is not problematic for installation in an injection molding machine. In many cases, molds are installed in the machine from above, between the tie bars and using overhead cranes. As long as the mold height is not greater than the height of the machine plates and as long as the width fits comfortably between the tie bars, everything is fine.

A layout in series is called an H-shaped runner , deriving its name from the shape of the runner for the inner four cavities. When more than four cavities are connected, the paths from the center of the mold are of different lengths. As a result, the melt reaches the outer cavities later, which is not good for the quality of the parts. Because the transition from the runner to the cavity is usually narrow, the melt flow slows down here and the melt might freeze off prematurely. In any case, the pressure profile of the melt over time will be different in the inner and outer cavities, so that the impression of the surface may not be uniform.

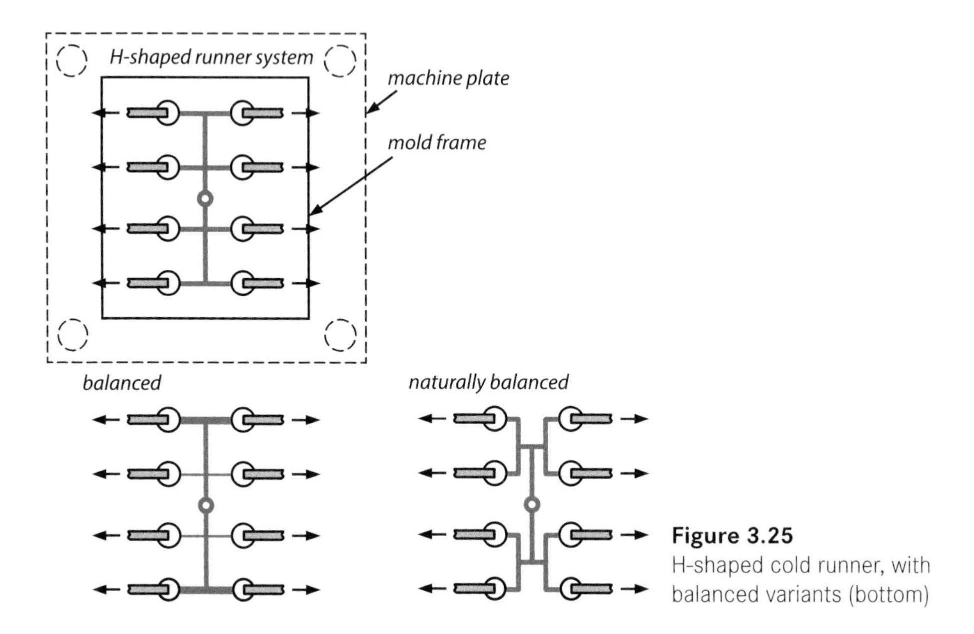

Figure 3.25
H-shaped cold runner, with balanced variants (bottom)

Balancing of the runner system for uniform cavity filling

Figure 3.25 shows possible ways to balance an H-shaped runner. The aim is basically to fill all cavities uniformly over time. Balancing occurs when the cross-sections of the runner are adjusted, such that the channels to the inner cavities initially become narrower. This increases the pressure requirement here, so that the melt flows more slowly here. Another possibility is to correct the gate (transition into the cavity). This optimization is done directly in production, with gates of cavities that fill too slowly being carefully ground larger.

By comparison, adjusting the flow path lengths is a natural balancing process. In this case, the runner weight and the pressure requirement increase considerably.

Flow cross-section of the runner

The shape of the runner influences the pressure requirement. A circular cross-section is ideal, because it offers the most favorable ratio of cross-sectional area to circumference (Figure 3.26). This means that the surface area for heat exchange with the mold is small. Circular cross-sections have to be machined into both mold halves, which is time-consuming and expensive. An additional disadvantage is that slide movements in the parting line could be impeded. For this reason, cold runners are predominantly incorporated into the ejector side only, which has the advantage that the cold runner can thus be reliably demolded via additional ejectors.

Common shapes for cold runners are trapezoidal or semicircular-trapezoidal (Figure 3.26, center), because the cold runner can be easily demolded via the inclined surface. Rectangular or semicircular cross-sections are unfavorable in view of the larger surface area.

The thickness of the cold runner depends initially on the component thickness s or the thickest component area. The channel diameter D is approx. 1.5 s_{max}.

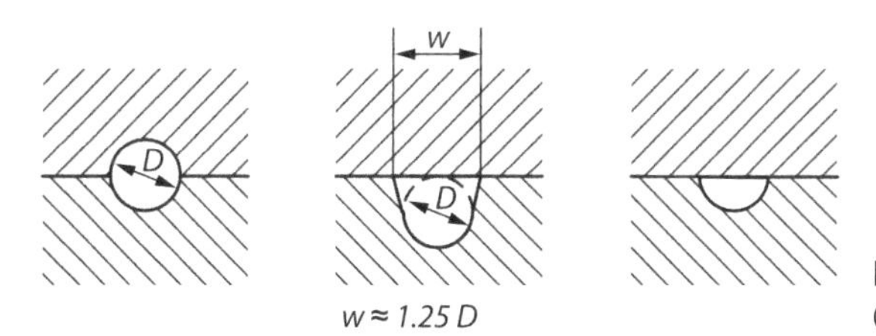

Figure 3.26
Cross-sections of cold runners

3.4.1.2 Gating to Cavities

The gate is the channel from the sprue or runner to the cavity

If mechanical reworking of each individual component is not planned or desired, the transition from the runner to the cavity represents a surface defect that is hard to avoid. With freedom from reworking in mind, direct gating via a conical sprue bar is ruled out and so the following requirements arise:

- The spruebar or runner should be easy to separate from the component, preferably by simply breaking it off.
- The gate should be as visually unobtrusive as possible.
- The cross-section of the gate should be as small as possible.

Possible locations for cold runner connections

Figure 3.27 shows useful gating options. Note that the gating is naturally in the same parting line as the gating system. This greatly limits the choice of gating to the part because, in multi-cavity applications, only the circumferential outer edge of the part can be considered. A tunnel gate dips below the parting line near the cavity and bonds the part on a surface perpendicular to the parting line. This gate type is self-separating. With the ejector movement, the sharp edge shears the runner system from the part. A tunnel gate can also be used to reach an auxiliary pin behind the actual component wall. This auxiliary pin is later broken off, so the visual defect of the gating is no longer on the visible surface of the component.

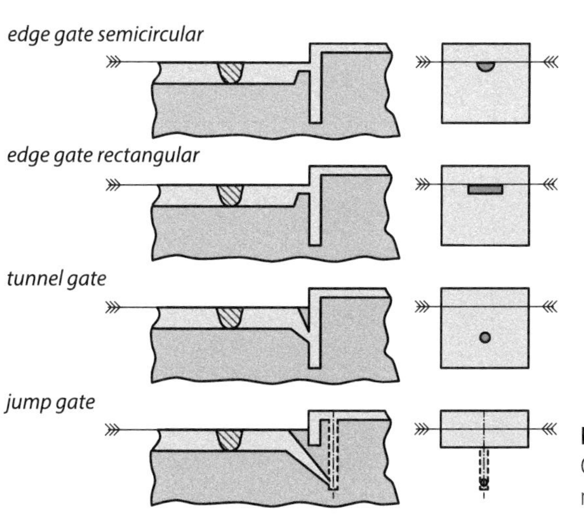

Figure 3.27
Gate shapes and corresponding markings on the component

Execution of a tunnel gate

A tunnel gate is executed with a stagnant space (conical bucket design) if possible (Figure 3.28). Here, the feed to the component is somewhat larger than the orifice itself. As a result, the tunnel gate remains molten for longer and thus allows a longer holding pressure phase to compensate for shrinkage. Another advantage of a stagnant space is that shear causes the flowing melt to be heated. Shear stress is always high when the melt flows rapidly in narrow channels, because the melt solidifies on the colder channel walls and shear forms between the flowing melt and the solidified non-flowing edge zone.

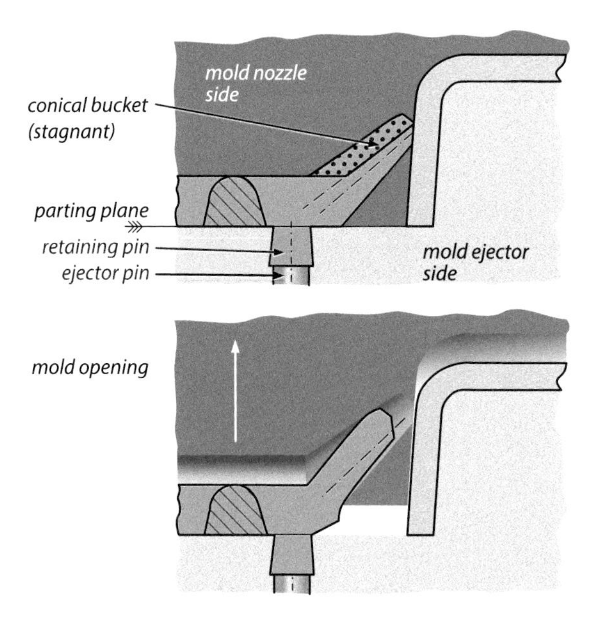

Figure 3.28 Tunnel gate with (dotted area) and without conical bucket

To ensure that a runner system with a tunnel gate can be demolded safely, a retaining pin with an ejector pin must be placed as close to the tunnel gate in the runner, because otherwise the runner system will be carried away when the mold is opened and the gate will not be separated safely.

Gate shape

Due to simple machining of just one half of the mold, a semicircle is initially the first cross-sectional shape to be considered (Figure 3.27). In many instances, however, the cross-section is enlarged and formed into a rectangle. In the case of very wide gates, this is also called a film gate. Very large widths have the advantage that the melt flows in uniformly over a surface area, whereby the orientations of plastic molecules and, in particular, filler fibers are more strongly aligned in the direction of flow. However, it should be borne in mind that a narrow cross-section can freeze more quickly. If, in the case of a very wide gate, the flow is not continuously equal at all points, the areas with slower flow may freeze, and then weld lines may be unintentionally formed in the part. Different flow velocities result when component areas behind the gate form a higher pressure resistance or are already filled after a certain time after the start of injection.

An additional parting line can also be used to gate a component on the surface of the nozzle side. This mold shape is referred to as a three-plate mold (Figure 3.29), and mold opening takes place via two parting lines, with one of the two lines opening in front of the second line via suitable mechanical elements. Inevitably, the gate will tear away from the component if the gate tapers towards the component. In this case, the gate is point-shaped. The runner system itself must be removed from the second parting plane via a suitable device before the mold closes again.

Cold runner gating on top of component with 3-plate mold

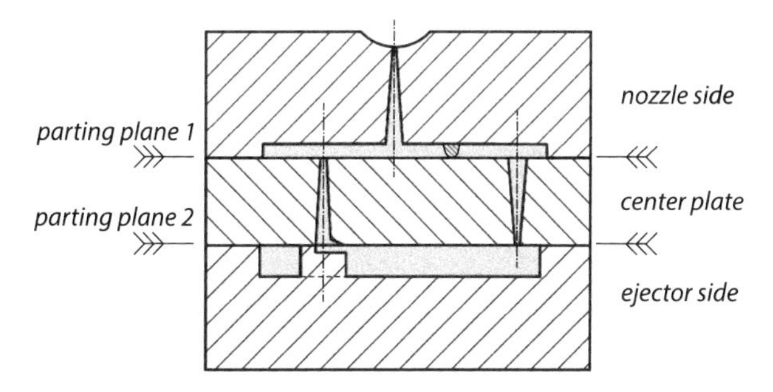

Figure 3.29 Gating via an additional parting line (three-plate mold)

If the draft angle increases the runner toward the component, the gate can be made at the inner edge of a cutout. The above-mentioned gate shapes, including the tunnel gate, can be selected here.

Although three-plate molds are costly because of the additional center plate and the mechanisms for controlling the opening movement, they offer greater design freedom:

- Several cavities can be gated, whereby the gate does not only have to be made via the component edge.
- The components can also be molded off-center.
- For larger component areas, several gating points can be used on one component; this prevents the filling pressures from coming within the range of the machine limit.
- It is possible to design the gating in the area of component openings. In this case, a second sprue bar would have to run conically from parting line 1 in the direction of the component opening in parting line 2 and meet an additional sub-runner. The connection to the component can then be made as usual via a gate in the parting line or via a sub-gate.

3.4.1.3 Demolding of the Runner System

Cold sprues bars can become very thick and thus cooling-time determinant

The cold runner system is usually located only in the ejector side ("B-side") of the mold. Sufficient ejector pins must be provided here to ensure that the runner system is reliably demolded. For the cold sprue bar a draft angle of 1° and a length of 100 mm results in an expansion of the diameter by approx. 3.5 mm. In extreme cases, the sprue bar can become so thick at the connection to the part that it solidifies only very slowly and may not be stiff enough for demolding and break off here. In this case, the sprue bar gets stuck in the nozzle side and causes a process malfunction.

Sprue bar retainer for safe demolding

To ensure that the sprue and runner system can be reliably demolded, mechanical locking is often provided (Figure 3.30). In many cases, this is done via a slightly conical undercut (top picture). During injection, this cavity takes up the material which is the first to leave the nozzle and which may be somewhat colder because of contact between the nozzle and the mold. The subsequent "correctly heated" material flows past this material. For this reason, the conical spigot opposite the runner is also called a cold slug catcher. This conical design is more suitable for tough and not too brittle materials.

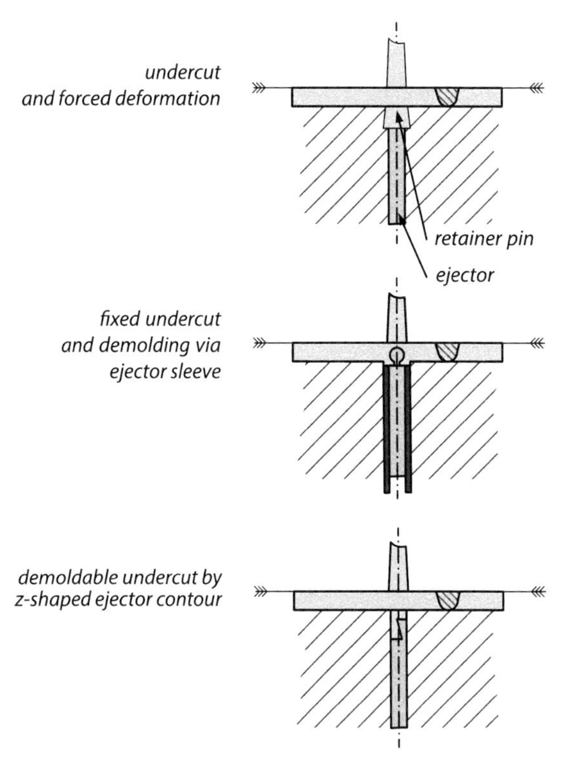

Figure 3.30 Demolding of cold runner

Alternatively, a fixed undercut can also be provided, which is formed by the end of a molded-on ejector. In this case, ejection takes place via an ejector sleeve. This design can be advantageous, especially for thick sprues, because the ejector pin is in the thick part of the sprue and thus reduces the wall thickness to be cooled here, making it less likely that the sprue bar will break off.

Retainers

Finally, the Z-pin is a good measure for forced demolding of the sprue bar. An ejector pin is ground in a Z-shape in the front area so that an undercut is formed. This undercut is released by the forward movement of the ejector. To prevent the sprue bar from sticking to the edge, the Z-pin should be secured against rotation in such a way that the extension of the undercut is directed downward when the mold is installed, so that the component can fall downward unhindered.

Cold runners lead to large mold opening widths

One disadvantage of cold sprues is the additional mold opening width, because the sprue must be removed with the component and this automatically increases the component height. As a rule, at least twice the component height must be provided for opening the machine, because the component must be removed past the cores (Figure 3.31). In the case of a bucket with a height of 300 mm, the complete bucket would just be visible with a mold opening of slightly more than 300 mm; demolding could then only happen into the nozzle side.

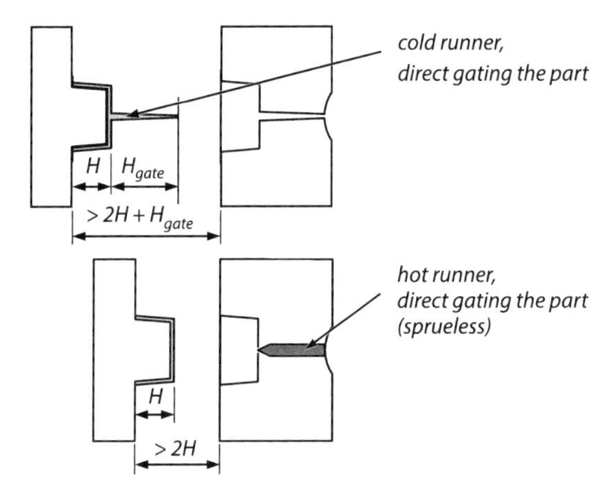

Figure 3.31 Necessary opening width for demolding components and, if necessary, sprues bars

Sprueless production thus facilitates demolding. There are three possible implementations, each of which prevents the melt from freezing completely prematurely:

- Mold with pre-chamber nozzle
- Insulating channel mold
- Hot runner mold

3.4.2 Mold with Pre-Chamber Nozzle

Thermal insulation using the pre-chamber nozzle as an example

Molds with pre-chamber nozzles (Figure 3.32) are hardly used anymore and will be discussed here only for further understanding. They are only suitable for relatively short distances between the actual machine nozzle and the cavity. A cavity is formed between the adjacent machine nozzle and the mold, which is filled with melt during the first cycle. Unlike the usual case, the machine nozzle does not lift off after each injection process, but locks off the cavity mentioned. The melt solidifies towards the mold edge, but, because of the low thermal conductivity of the plastic melt, a hot plastic melt plug remains in the center area. By design, an air gap around the actual pre-chamber nozzle can be provided as an insulator to prevent the entire melt volume from freezing off too quickly. Cooling of the pre-chamber nozzle in the front area is intended to ensure that the transition to the cavity freezes here and that no troublesome filaments are drawn during demolding.

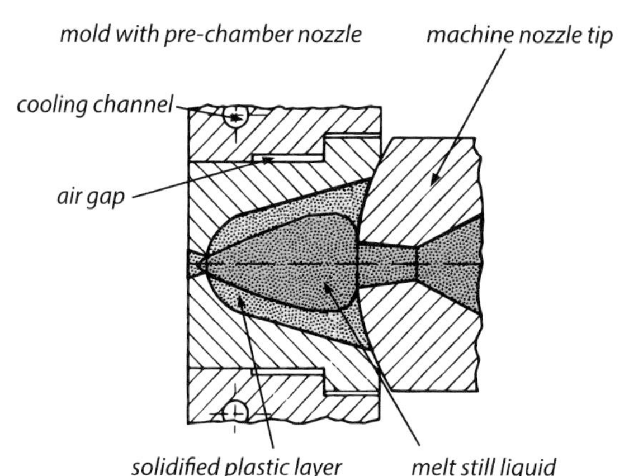

Figure 3.32
Pre-chamber nozzle for sprueless injection molding [image source: Kunststoffverarbeitung im Gespräch, BASF, 1979]

One disadvantage of pre-chamber nozzles is the large melt volume. Because the amount of melt for each cycle is precisely matched to the cavity, the cavity is not completely filled, at least in the first cycle. Thus, starting up the mold after a malfunction is not unproblematic.

Heated pre-chamber nozzle (hot runner prototype)

The use of highly thermally conductive nozzle tips improves this circumstance (Figure 3.33). Due to the good thermal conductivity, the temperature difference within the nozzle tip is very low and thus the melt can be kept very close to liquid and hot until just before the cavity. Here, too, the nozzle remains permanently in contact with the mold. The cavity between the nozzle tip and the mold is filled with melt the first time, which then solidifies for the most part and largely insulates the hot nozzle tip from the cold mold. Some of the plastic material thermally degrades over time, but this is not problematic because only the melt in the area of the opening to the cavity actually enters the mold cavity upon injection. This happens with every cycle, so there is no risk of thermally damaged material appearing on the part surface.

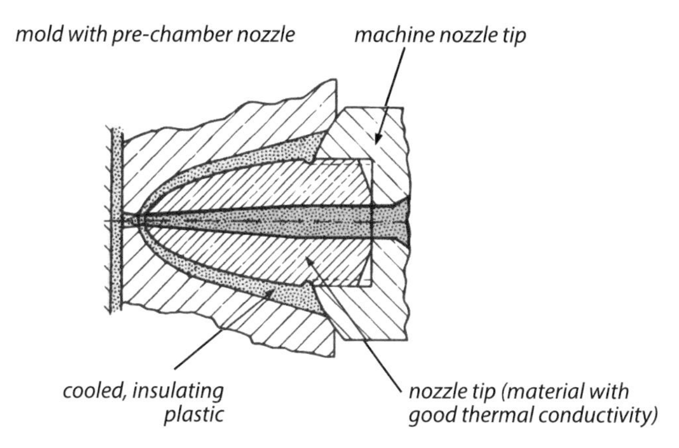

Figure 3.33 Heated pre-chamber nozzle for sprueless injection molding [image source: Kunststoffverarbeitung im Gespräch, BASF, 1979]

One disadvantage of pre-chamber nozzles is the higher nozzle-contact pressure required. Due to the large melt diameter in the direction of the nozzle and the high possible injection pressures – a pressure of approx. 2000 bar can prevail in the nozzle area – high forces result which must be compensated by the contact pressure forces of the machine nozzles so that the injection unit is not pressed back.

Pre-chamber nozzles with heated nozzle tips are mainly used in combination with hot runners; here, cyclical lifting of the nozzle is undesirable because of possible material leakage from the hot runner.

3.4.3 Insulating Channels

Insulating channels keep the melt liquid

In principle, an insulating channel is an extended pre-chamber nozzle and has the same advantages and disadvantages. The melt remains plastic toward the center if the diameter of the channel is large enough (Figure 3.34). Provision of a measure of insulation via air gaps is not possible. The main problem of insulating channels is the start-up process, because, if the chosen diameter is very large, the insulating channel takes up to two shots of melt. It is therefore necessary to inject up to twice before melt enters the cavities.

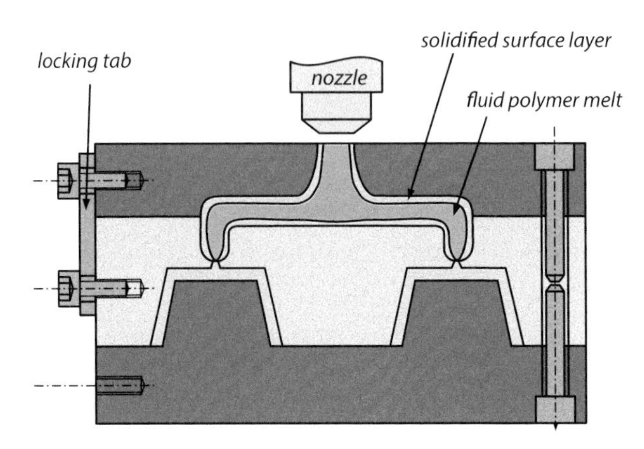

Figure 3.34 Insulating channel for sprueless injection molding

After longer malfunction periods, the entire insulating channel may be frozen, in which case the cooled plastic must be removed. For this purpose, the level of the insulating channel must be designed as a second parting plane, which is kept closed with a locking tab for cyclical operation. The second parting line can be opened by folding down the locking tab.

Insulation channel diameter

Insulating channels are hardly ever used because it is difficult to find a compromise with regard to the diameter:

- Large diameter → channel freezes slowly and so the system becomes insensitive to possible disturbance times during demolding.
- Small diameter → easier start-up and larger available melt volume due to lower compression volume.

Issue: compression volume

If the diameter is very large, the channel may contain too much compressible melt, and so the maximum possible shot volume of the screw is no longer sufficient and a larger injection unit becomes necessary. The shot volume of the injection unit is calculated from the stroke of the screw and the screw diameter; it should be approx. 20% larger than the volume of the cavity, because the hot melt shrinks during cooling, to an extent depending on the type of plastic, and in addition it must be taken into account that the melt is compressed by approx. 10% at an injection pressure of 2000 bar. Now, if the insulating channel is very large and corresponds, for example, to twice the shot volume, the compressed volume increases to 30% (Figure 3.35). The compression volume cannot enter the mold. In a sense, it is not effective volume, and is only necessary for pressure buildup.

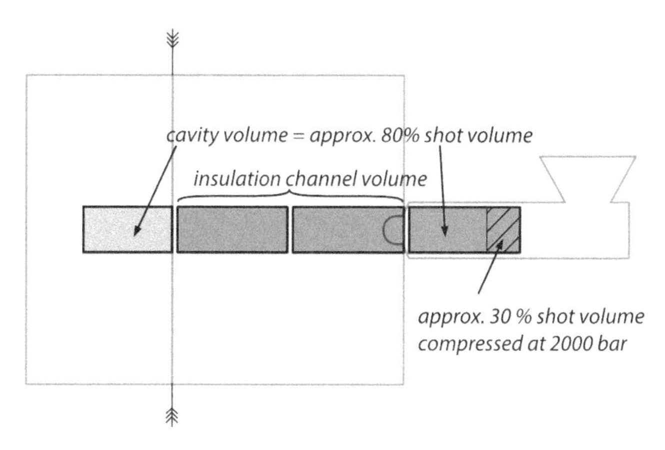

Figure 3.35 Compression volume of an insulating channel

3.4.4 Hot Runners

Advantages of hot runners

The disadvantages of insulating channels with regard to the difficulty of controlling freezing of the melt can be remedied with electrical heating. Despite the additional expense for this heating in the form of heating elements and temperature regulation, hot runner systems are largely standard in injection molding machines. They offer various advantages and possibilities:

- *No sprue material:* The entire melt distribution system in the mold is heated, so there is no superfluous waste that has to be returned to the process. Sprue and runner recycling requires a comminution mill, which is usually not located directly at the machine. In the case of central material preparation with sprue or scrap part mills, there is a risk of material contamination.
- *Shorter mold opening travel:* The demolding of the sprue system is problematic. For a smooth cyclical process, the sprue must be safely demolded. The mold must also be opened by the length of the sprue so that the component can be safely demolded (Figure 3.31). The opening width of the mold is ultimately limited by the machine size.
- *Reduced pressure losses:* A cold runner increases the flow-path length, thus increasing the pressure requirement. With hot runners, this pressure requirement can be reduced so that higher pressures become available for the cavity filling process and for reproducing the surface texture of the cavity.
- *Better holding pressure effect:* In non-heated systems, the gate freezes quickly during the holding pressure phase. Due to heating of the nozzles in the transition to the cavity, the holding pressure can act for longer, and this has a positive effect on shrinkage and warpage.

- *Better gate optics:* Depending on the system, the gates can be very clean. With valve gate systems, only an impression similar to an ejector pin may remain visible on the surface.
- *Stack and tandem molds:* Molds with multiple parting lines for components can be used to increase performance.
- *Gating from the inside is possible:* This is especially interesting for decorated containers, e.g. magarine pots. If a film with a printed image for the outside of the component is to be inserted into the mold and back-injected, gating from the outside is not possible. Depending on the container height, that would result in a large gate length that could not be easily bridged with any type of pre-chamber nozzle.
- *Cascade technology possible:* Particularly for elongated components, several gating points are necessary because of the possible flow-path lengths. This automatically results in weld lines between the gates. This is referred to as cascade technology if the various gates can be opened independently in chronological sequence via appropriate mechanisms. If a nozzle is opened only after the melt has flowed over it from an upstream nozzle, this inevitably results in no weld lines (see Section 3.4.4.4).

Hot runners consist of sprue bushing, hot manifold and hot drops (nozzles) (Figure 3.36). These elements are installed in the nozzle side (A side) of a mold with as little direct contact as possible, because as little heat as possible should be exchanged between the hot runner and the mold. In principle, the most suitable insulator is an air gap (Figure 3.37). The manifold is in simple cases merely an elongated bar, but it can also have various branches.

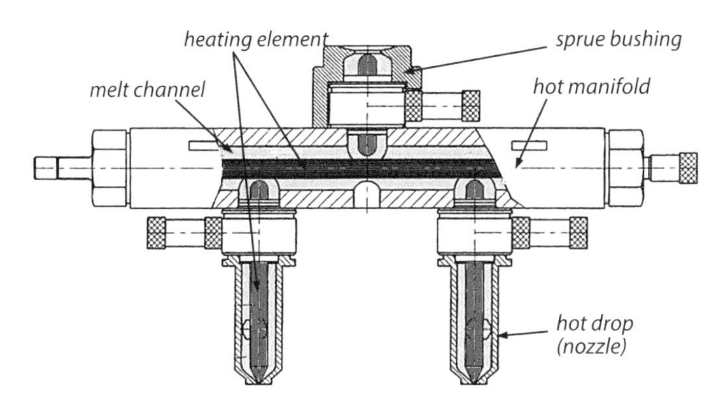

Figure 3.36 Hot runner, internally heated [image source: Mennig, "Werkzeuge für die Kunststoffverarbeitung", Carl Hanser]

Due to thermal expansion, the manifold is shorter in the cold (assembly) state than in the heated state. Therefore, there are two variants for the connection of thehot drops to the beam:

Thermal expansion of hot manifolds

- In the sliding press connection, thehot drops lie flat against the manifold. The distance between the hothot drops is determined by the dimension between the various gates and does not change when the hot runner is heated up. The hot manifold itself will expand while heating. When heated, the spacing of the orifices in the manifold matches the pitch of the gates. The hot drops grow longer as a result of heating and thus generate a pressing force on the manifold so that no melt from the hot runner enters the space between the hot runner and the mold.

Sealing of the hot drop (nozzle) via press connection

- Screwed-in hot drops are inevitably bent along their length due to the thermal expansion of the manifold. Therefore, either small manifold lengths must be present here or the hot drops must be relatively long. As an approximation, the length of screw-in hot drops must be at least twice as long as their distance from the center of the manifold. For example, a hot drop with a distance of 150 mm from the center of the mold should be at least 300 mm long.

3.4.4.1 Internally Heated Systems

In internally heated systems, electric cartridge heaters are used for heating, around which the molten plastic flows (Figure 3.36). On the outside, these hot runners are largely cold, despite an insulating gap to the mold. For this reason, a solidified or at least a very tough edge layer forms on the channel wall. Two *advantages* result:

Internally heated hot runners cause little heating of the surrounding mold

- The tough surface layer seals the system very well and so leaks are unlikely when using sliding hot drops.
- Due to the low outside temperature, the heat input in the direction of the mold is low and so the demand for electrical power is also lower.

The *disadvantages* of internally heated systems are:

- Color or material changes are particularly problematic and cannot be carried out quickly without complete mechanical cleaning.
- The pressure resistance is quite high if the chosen channel diameter is not large enough. With large diameters, however, the problem of compression volume arises (see Figure 3.35).

3.4.4.2 Externally Heated Systems

Externally heated systems have a lower pressure requirement and a lower melt volume. The melt flow channel is completely hot, there is no solidified layer preventing possible leakage.

Externally heated hot runners have a completely hot melt channel

Flexible heating conductors embedded in grooves on the outside of the manifold are commonly used. Figure 3.37 shows an alternative heating system with installed flat heating plates and insulation gaps, the pressure disks for sealing, and the centering of the system. The manifold is fixed over the center of the mold and can expand along its length to both sides. Accordingly, the lateral centering engage with slotted holes and align, for example, the vertical axis in the installed condition.

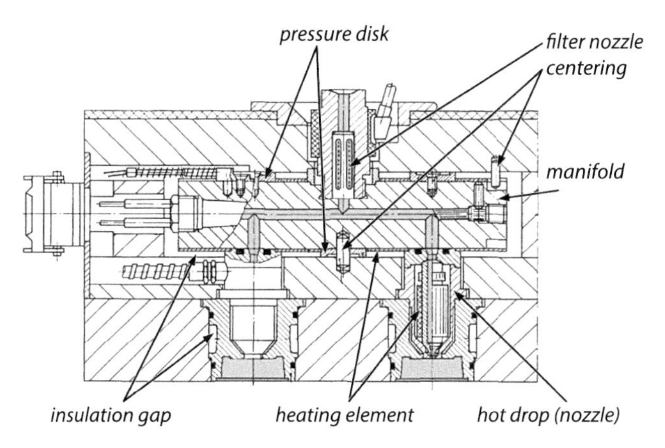

Figure 3.37 Hot runner, externally heated [according to Mennig, "Werkzeuge für die Kunststoffverarbeitung", Carl Hanser]

Because there is no solidified edge layer as in the internally heated system and because the hot drops cannot be screwed tightly due to the linear expansion of the hot runner manifold, the hot drops must be sealed by surface pressure in the heated state. For this purpose, pressure disks are required opposite the nozzles. These pressure disks should transfer as little heat as possible between the hot runner manifold and the mold. For low heat transfer, they are either made of a material that is a poor heat conductor or they have grooves in their surface to reduce the heat-transfer area.

3.4.4.3 Hot Runner Nozzles

An important requirement for hot runners is thermal separation from the mold. This is particularly difficult in the region between the gate and the cavity, because here a large temperature gradient has to be made possible over very short distances.

There are three types of nozzles:

- Open nozzle
- Nozzle with hot tip
- Valve-gate nozzle

An open nozzle (Figure 3.38) is, in principle, an ante-chamber nozzle (see Figure 3.32). The opening cross-section is relatively large, so that wear is also quite low in the case of plastics with abrasive fillers. A disadvantage is a strong and often unclean marking on the component in the form of a plastic tear-off that is easily noticeable and irregular from cycle to cycle. In this case, it is advisable to form a calotte on the component. This refers to an indentation around the tear-off point that accommodates the sprue slug. In many cases, for example, a container is gated at the bottom area. The indentation of the surface ensures that the container bottom later rests well on a flat base and that the component does not tilt on the break-off material.

Open nozzles for viscous, filled plastics

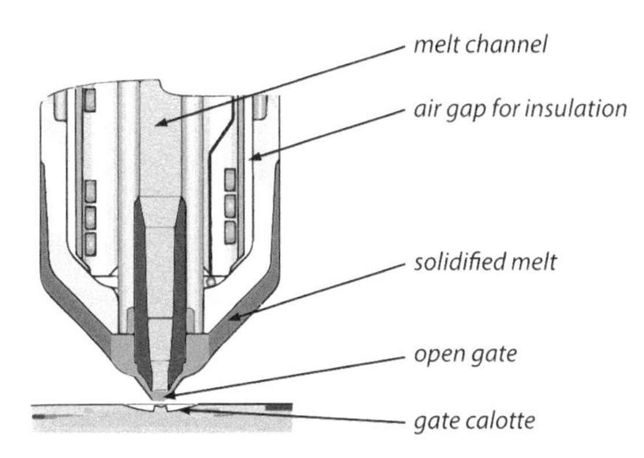

Figure 3.38
Hot runner nozzle, open version [image source: Günter Heißkanaltechnik, Frankenberg]

Semi-crystalline plastics can solidify quickly in the area of thermal separation between the nozzle and the cavity, causing process disturbances. In that case, heated tips can be used for the open nozzles (Figure 3.39). Here, the melt must flow through an insert and at the front end via a lateral opening into the actual gate area. This inevitably results in weld lines, which leave visible linear marks, especially when color pigments are used.

Open nozzles with tip for fast-solidifying plastics

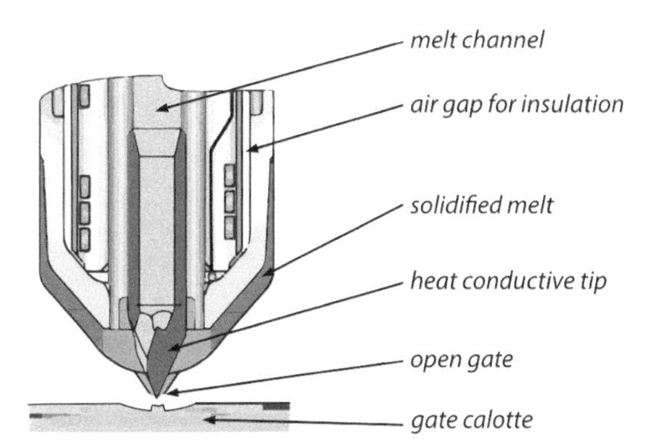

Figure 3.39
Hot runner nozzle, version with tip [image source: Günter Heißkanaltechnik, Frankenberg]

Actuated valve gate nozzles and their advantages

Hot runner nozzles can be closed with externally actuated needles (Figure 3.40). This has several advantages:

- Some plastics make for very fluid melts and tend to leak out of the hot runner when the part is demolded. As long as the component is still in the cavity, it seals the nozzle gate itself.
- Occasionally, plastics containing blowing agents are used, e.g. if the plastic is to be foamed to reduce density and thus save weight. This creates a pressure of about 30 bar in the hot melt, which would drive the plastic out of the hot nozzle when the mold is opened. Valve gate nozzles prevent this.
- With very short cycle times, which is the case with thin-walled packaging, the machine still has to generate melt while the mold is already opening and demolding the component. This creates considerable pressure in the screw antechamber, which propagates into the hot runner. The only way to prevent the melt from leaking is to use valve gate nozzles.
- Open hot runner nozzles do not always leave the same melt break, usually in the form of a slight elevation. With shut-off nozzles, this mark is leveled. Depending on the quality of the valve gate system, an impression remains which is very similar to that of an ejector.

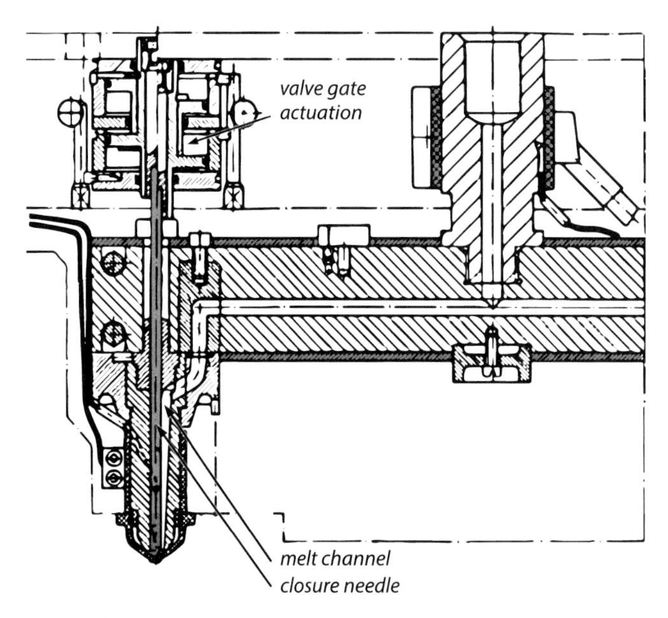

Figure 3.40 Valve gate nozzle of a hot runner

Valve gate system drive

The valve gate systems are complex and require additional installation space for the needle actuation. This needle actuation must be thermally insulated from the hot runner itself. Pneumatic actuation is the best solution, since heating of the actuation system by the hot runner is less critical than with hydraulic actuation. The disadvantage of pneumatic systems is the poor fine-tuning of the speed. Electric drives are very advantageous here, but they are also quite expensive.

A filter nozzle is optional for hot runners. Basically, there is always a risk of impurities entering the machine via the material supply. These impurities can clog narrow nozzle gaps and cause malfunctions. Filter nozzles keep impurities in front of the hot runner and facilitate maintenance in the event of a malfunction.

Filter nozzle protect hot runner systems from contamination

3.4.4.4 Cascade Technology

For large or very elongated components, it is often necessary to define several gating points if the flow path/wall thickness ratio of 150 : 1 is exceeded by far. This would inevitably result in weld lines when the melt streams from different nozzles come together.

Hot runner cascade opens several nozzles with a time delay

With independently controllable hot-runner valve-gate nozzles, these can now be opened at staggered intervals, so that a second nozzle only allows melt into the cavity when the flow front has overflowed this gate by approx. 20 mm (Figure 3.41). This is a good way to avoid weld lines arising from the merging of two melt streams.

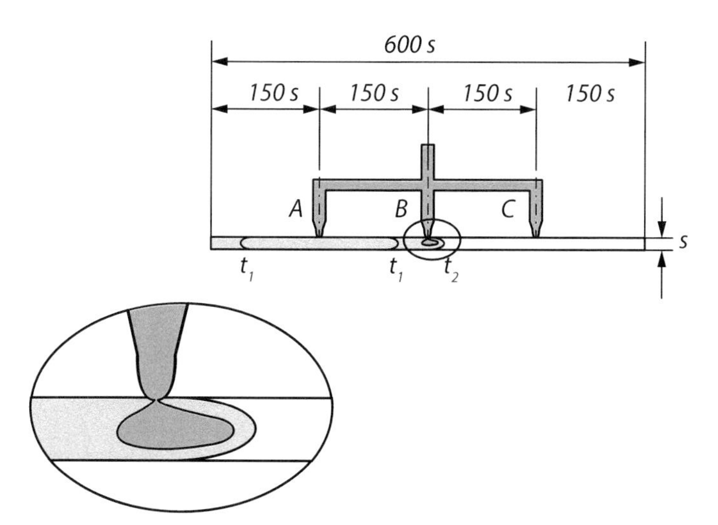

Figure 3.41 Hot runner cascade technology with three independent shut-off nozzles

The example shows a part with a length of 600 times the wall thickness, which would be too large for a realistic flow-path-to-wall-thickness ratio of 150 : 1. The part could be produced with two nozzles (A and C), but this would inevitably result in a weld line at location B. With three valve gate nozzles, this part could be produced without a weld line. When selecting the valve gate nozzles, care should be

Hot runner nozzles open only after the melt has flowed over them.

taken to ensure that the opening speed of the valve gate can be controlled. In the hot runner manifold, pressure initially builds up in front of the still closed nozzles B and C as long as melt is only flowing through nozzle A. The pressure in the hot runner can be adjusted by opening a further nozzle. With the opening a second nozzle close behind the flow front, the pressure is released, with the result that the melt flow front a undergoes a certain degree of acceleration. This can result in shadow-like markings on the component surface. If the nozzles are opened somewhat more slowly, this surface defect can be readily avoided.

3.5 Temperature Control

Goals of temperature control

The temperature of the injection mold in the area of the cavity has an influence on:

- Cycle time
- Quality of the surface of injection molded parts
- Warpage and dimension of injection molded parts

The cooler the mold, the faster the cycle

For short cycles, the mold should be as cold as possible. The higher the temperature difference, the higher the heat flow. The heat of the molten plastic must be dissipated to the mold as quickly as possible, and so low temperatures are advantageous. Roughly put, a mold temperature lowered by 10 °C shortens the cooling time by approx. 10%.

Warmer molds enable better surface quality

For a good surface impression, the mold should be as hot as possible. Immediately upon first mold contact, the melt flow forms a solidified edge layer which is so tough that it cannot reproduce small depressions in the surface. The warmer the mold, the better the plastic can reproduce the surface. The temperature level depends on the grade of plastic, any added fillers and the type of surface texture.

Cooling rate influences possible crystallization

The dimensions of injection molded parts are also influenced by the mold temperature. An injection molded part is not completely cooled at the time of demolding; on the contrary, it is still quite hot inside. As it continues to cool without mold contact, the warmer areas of the plastic will still contract somewhat. Depending on the structure and rigidity of the injection molded part, warpage may now occur. This warpage is often observed on the inner flanks of flat box-shaped components (see Figure 1.37). In the case of semi-crystalline plastics, temperature can also influence crystallization. With cold molds, heat is quickly removed from the plastic, and so the plastic cannot crystallize as well. As crystallization increases density and reduces volume, less shrinkage can be expected with cold molds. In principle, a plastic can continue to crystallize for a long time after production, especially if it is exposed to higher temperatures. In that case, the component may still warp later.

The temperature ultimately selected for production depends on the type of plastic; engineering plastics (PA, ABS, PC, ...) in particular should be processed at high mold temperatures.

The mold is a heat exchanger

The temperature of the mold will be set after approx. 20 cycles in cyclical operation; it is the result of different heat fluxes ($\dot{Q}$) which are supplied or dissipated (Figure 3.42). Whether heat is supplied or dissipated by the temperature control medium ($\dot{Q}_{medium}$) depends strongly on the total mold temperature. At low temperatures, convection and radiation ($\dot{Q}_{convection/radiation}$) are almost negligible; above approx. 80 °C, a large amount of heat is dissipated into the environment. Then the temperature control medium will bring additional heat into the mold, if necessary, so that the temperature required for the quality can be maintained.

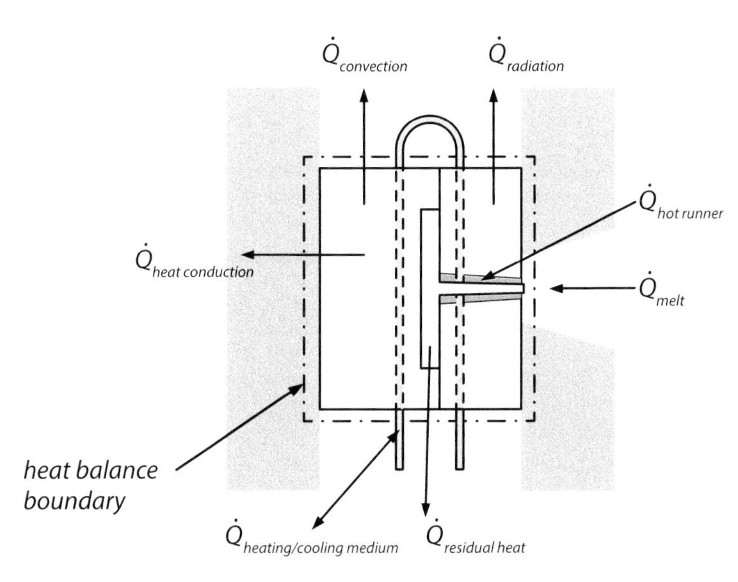

Figure 3.42 Heat balance of an injection mold

Mold temperatures increase within the first 20 cycles

The mold temperature is not constant, as the heat of the melt is supplied cyclically. At the start of production, the mold has the temperature of the temperature control medium. Within approx. 20 cycles, a slightly higher actual temperature is established in the cavity area. Figure 3.43 schematically shows the change in temperature of the cavity surface. Immediately upon contact with the melt, a temperature rise occurs and, in the holding pressure and cooling phase, the temperature equalizes again. In an overall uniform cycle, the lower temperature of this oscillation is the minimum temperature required for quality. If this temperature does not change significantly during cyclical operation, it is referred to as a quasi-steady state.

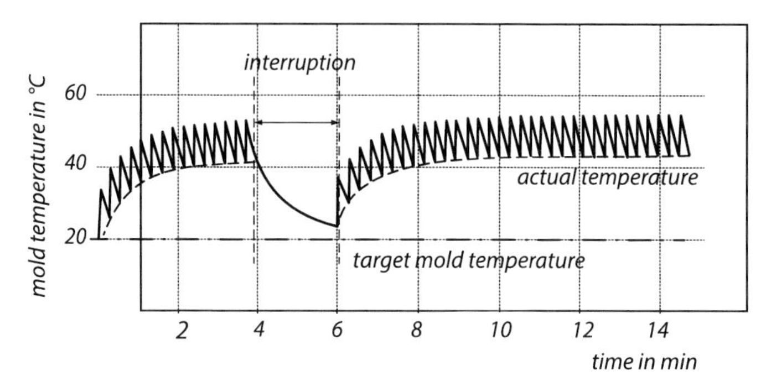

Figure 3.43 Temperature of a mold during start-up or after a process interruption

During the cycle, the temperature fluctuates by up to 10 °C

The temporal representation of the temperature differences refers to only one location in the mold. Due to the limited thermal conductivity of the mold steel, the temperature oscillation is greatest directly at the cavity surface and negligible directly at the cooling channel (Figure 3.44). The magnitude of the cyclical temperature difference at the surface depends on the temperature difference between the melt and the mold.

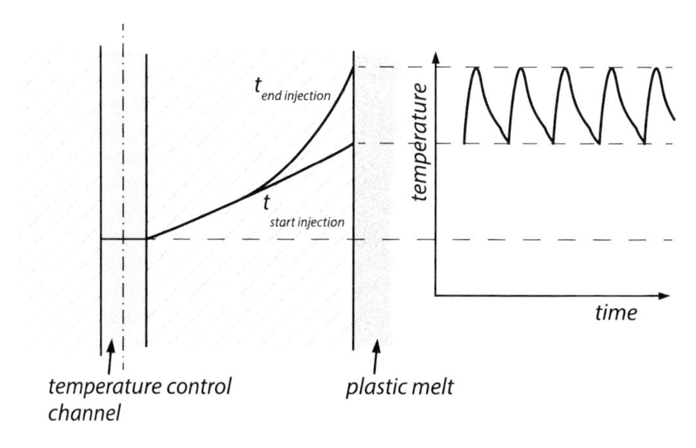

Figure 3.44 Temperature differences during cyclical operation in an injection mold [image source: Thienel]

Actual and setpoint temperatures deviate from each other

At the start of the injection process, a defined minimum mold temperature must prevail, which is the lower value in the cyclical temperature oscillation. This actual temperature is decisive for the quality of the surface. It does not match the set target temperature, because a temperature difference is necessary for heat conduction. Therefore, the temperature at the cavity surface is always higher than the temperature of the temperature control channel during cyclical operation. Theoretically, it would be possible to readjust the temperature of the temperature control medium so that the actual temperature and the set temperature are the same. In practice, however, it is hardly possible to have the desired set temperature at every point on the cavity surface, because the position of the cooling channels generally does not permit optimum temperature control at all. It should therefore be borne in mind that, after the process start, a few cycles will elapse before the mold is at the right temperature for the desired surface quality.

As soon as the melt makes contact with the mold during injection, there is a short temperature peak of up to 10 °C, and with the end of injection no further heat is introduced into the mold, so that the temperature drops again. With increasing distance to the cavity wall, this temperature peak becomes smaller, and constant conditions prevail directly in the area of the temperature control channel, plotted against time.

Temperature differences of 1 °C/mm are possible

Figure 3.45 shows the difference between a mold steel and an alloy of copper-beryllium of very good thermal conductivity. Here it is clear that, with a poorly thermally conductive steel, local temperature differences of approx. 1 °C/mm can prevail in cyclical operation. It is also clear that the temperature of the temperature-control medium must be changed for a more thermally conductive material for a defined mold temperature. If the mold is to be used at a defined higher temperature, the temperatures of the temperature-control medium must therefore be set higher when better-conducting steels are employed.

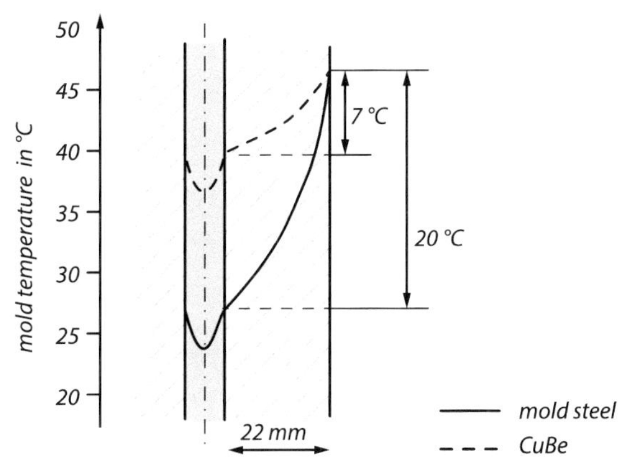

Figure 3.45
Influence of mold steel on local temperature differences in the mold [image source: Wübken]

Temperature control close to the contour does not shorten the cycle time

It is a widespread misconception that faster heat conduction in the mold through cooling channels located close to the contour or materials with higher heat conductivity will shorten cooling times. The thermal conductivity of plastic is considerably lower than that of mold steel. The dissipation of heat from the interior of the component is therefore a physical limit that cannot be shifted unless the thermal conductivity of the plastic itself is changed, e. g. via fillers.

What is important is uniformity of temperatures at the cavity surface. If large differences occur here, because heat cannot be dissipated in parts of the mold, the cycles will be prolonged. From this point of view, mold materials with good thermal conductivity are very useful for reducing the temperature differences between cooling channels at the cavity surface.

Effectiveness of temperature control increases with turbulent flow

With regard to the temperature control medium, the figure shows different temperatures across the cross-section of the temperature control medium channel. This is the case when the temperature control medium flows laminarly. Under turbulent flow, mixing of the temperature control medium occurs along with equal temperatures across the cross-section of the channel. Where channel cross-sections are below approx. 3 mm and flow velocities are low, turbulent flow is rather unlikely, and the effectiveness of the temperature control decreases.

Steels of higher thermal conductivity lead to lower temperature differences on the surface

Local temperature differences are also present at the surface if the cooling channels have a wide spacing B and if the distance s to the cavity surface is small (Figure 3.46). The mold steel is a heat-flow resistor, and the smaller the distance between the cooling channels and the cavity surface, the greater are the local temperature differences. This is referred to as cooling error, but there are no generally valid statements as to which temperature differences are permissible. If the temperature differences are very high, the surface texture may be reproduced differently, i.e. the gloss effect of the surface will not be uniform. In addition, the time required until demolding may be longer.

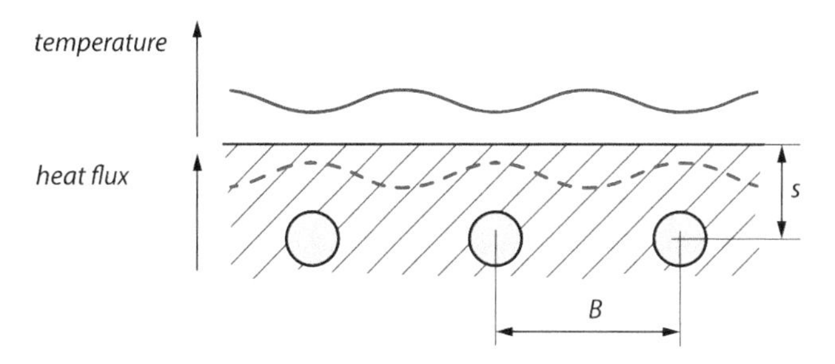

Figure 3.46 Cooling errors are non-permissible temperature differences at the cavity surface

This information has the following consequences for the design and development of a mold:

1. A component design with a fissured surface, e.g. a very narrow ribbing, will result in larger temperature differences. Possibly this will negatively influence the expected cycle time and thus the calculated production costs.
2. A flat component surface will have a lower cooling error:
   - The distance of the cooling channels should tend to be further away from the cavity wall and should be in the order of magnitude of the distance of the cooling channels from each other ($B \approx s$). In general, temperature control as close

as possible below the surface is recommended, so that cooling takes place quickly. With a view to uniformity of the surface temperature, however, it makes more sense to increase the channel spacing and to reduce the temperature of the temperature control medium.

- If narrow cooling channel distances are not possible for geometric reasons, a mold insert made of a better heat-conducting material can reduce the cooling error.

3. If the mold is to be operated at high temperatures, the cooling channels should not be located close to the contour. If they are, in addition to the above-mentioned temperature control errors, the control oscillations of a temperature control unit would generate cyclical temperature fluctuations. With a larger distance between the temperature control channels, the mold becomes somewhat more thermally inert, changes in temperature become slower and weaker, and, from the point of view of consistent quality of the injection molded parts, this is better.
4. If the mold is to be operated as cold as possible, cooling channels as close as possible below the cavity surface are better. In addition, material with good thermal conductivity should be used as far as possible for the mold inserts.

3.5.1 Concepts for Temperature Control

A distinction is made in the temperature control of injection molds between:

1. Continuous flow temperature control
2. Impulse cooling or discontinuous temperature control
3. Variotherm concepts or intermittent temperature control

The designer should be aware of these concepts and the specific objectives, because they can partially resolve the difficulties described above.

3.5.1.1 Continuous Flow Temperature Control

Flow-through temperature control: inlet and outlet temperature should be the same

The temperature control medium flows through the mold at a constant speed. It should be noted that the temperature control medium is not noticeably heated on its way through the mold, so that the mold itself is not warmer in the area of the temperature control medium outlet than in the area of the temperature control medium inlet. A small temperature increase is achieved by the following:

- The volume flow of the temperature control medium should be large. Cross-sectional narrowings should be avoided as far as possible.
- Higher pumping capacities of the temperature control units may be necessary.

- The temperature control channel length within the mold should not be too long. In production operation, several temperature control channels are often connected in series by means of hose bridges, which is not helpful with regard to the temperature increase.

Temperature control in the mold should be done in series

A temperature control channel in the mold can certainly control the temperature of several neighboring cavities, especially if there is not enough space for independent temperature control channels. In this case, the temperature control channel should supply the cavities in series one after the other (Figure 3.47). When several cores of a temperature control circuit are connected in parallel, malfunctions during operation are very difficult to detect. If, for example, the flow at a cavity is blocked – this can be due to calcification, rust, or contamination of the temperature control medium – this cavity becomes warmer and warmer, although the process may not necessarily be disturbed or interrupted. Quality defects (surface, dimension) are then only noticed at a very late stage, possibly when the parts are already packed or at the customer's premises. In a series connection, a flow blockage leads to a visible disturbance, because many components are directly affected.

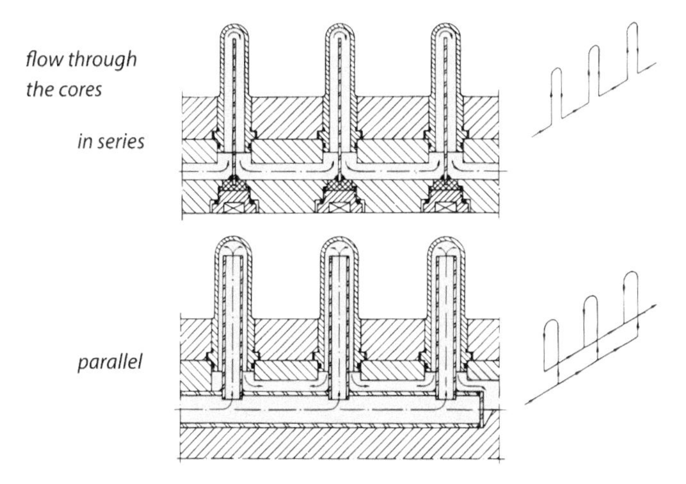

Figure 3.47 Flow temperature control; above: in series; below: in parallel [image source: Mörwald]

Always flow to several circuits in parallel (individually)

If several circuits are present on the mold, the flow should always occur in parallel. The flow of each individual circuit can be monitored by simple means. Due to the parallel flow, the circuits remain short, so that any noticeable temperature increase in the temperature control medium is low.

3.5.1.2 Pulse Cooling/Discontinuous Temperature Control

Temperature control in the case of non-continuous flow of temperature control medium

The heat flux dissipated via the temperature control medium is:

$$\dot{Q}_{medium} = \dot{m} c_p \Delta T \tag{3.2}$$

where the quantity of temperature control medium of specific heat capacity (c_p) flowing for a certain time ($\dot{m}$) has a certain temperature difference ΔT. In principle, it is possible to time the flow and thus reduce the quantity of temperature control medium. If the temperature of the temperature control medium is reduced at the same time, the heat flow remains the same.

This type of temperature control is used for two different purposes.

Reduced outlay on the periphery

Outlay on the periphery is reduced by molds that are kept at a higher temperature level. In most cases, the mold halves or the respective temperature control circuits are run at different temperatures. This is necessary because different mold areas will also have different temperatures after the start-up process. For example, if a core is difficult to cool, it is supplied with the temperature control medium at a lower temperature to prevent it from becoming unacceptably hot. Under continuous flow temperature control, as many temperature control units are required as different temperatures are desired. Under discontinuous temperature control, only a corresponding number of switching valves are required, which can control the heat quantity via the pulse length. In addition, energy can be saved because the temperature control medium does not necessarily have to be heated to a higher temperature. In principle, a cold temperature control medium at a temperature of 20 °C can enable defined mold temperatures of 80 °C, for example, in this way.

Cooling with refrigerant pulses

Very good cooling of slim cores is possible if the temperature control medium undergoes a phase transition from liquid to gas. Here, the principle of refrigerators is used. A suitable temperature-control medium is compressed to such an extent that it is liquid at room temperature. This liquid is fed under pressure into a core and expanded there, where it evaporates and extracts heat from the environment, i.e. the mold. CO_2, for example, is a refrigerant which is released into the environment after evaporation. Furthermore, common air-conditioning refrigerants are used, but are kept in a closed circuit and liquefied again by a compressor. With these refrigerants, local mold temperatures of below 0 °C are possible. Given correspondingly short pulses, the quantity can be measured very well, allowing the temperature to be controlled in areas where it is otherwise hard to control.

3.5.1.3 Variothermal or Intermittent Temperature Control

Cooling with alternating high and low temperatures

Molds should be as cold as possible in the cycle for rapid melt cooling and as hot as possible for surface reproduction. This contradiction is solved by variothermal temperature control, which offers a number of advantages:

- Visible weld lines can disappear completely if the mold temperature at the time of first melt contact is approximately as high as the glass transition temperature of the plastic material.
- Fiber-reinforced plastics form a very smooth surface. Usually, fiber-reinforced plastic parts have a rough and uneven surface.
- Highly polished mold surfaces are also reproduced so well with ABS that the component surface does not need to be subsequently painted. This only applies to the gloss; for scratch protection, additional painting would still be necessary, although further pretreatments can largely be omitted.
- Very fine textures in the micro range with a large width-height ratio can only be reproduced with mold temperatures in the range of the melt temperature.

Variothermal temperature control is energy-intensive

Basically, variothermal temperature control is energy-intensive, because some parts of the mold have to be actively heated and cooled again. There are different approaches to this, which differ greatly in terms of technical effort and energy requirements.

Table 3.2 Concepts of Variothermal Temperature Control

Type	Technical Effort	Energy Demand
Two-loop technology	++	–
Resistance heating	+	+
Induction heating	–	++

Approaches using two loops

In the two-loop technique, temperature control media of different temperatures are passed alternately through the temperature control channel. Depending on the temperature control unit, the warm medium may not be separated sufficiently well from the cold medium. Two independent temperature control circuits are usually not possible for reasons of space. The temperature change of the mold is relatively slow and may well lead to longer cycle times. Relatively large areas of the mold go through the temperature change, so that the energy requirement for this becomes very high.

Electric heating and water cooling

With electric resistance heating, the technical effort is somewhat greater. In the case of large-surface cavity contours, however, it is possible to insulate the electric heater from the mold so that largely only the areas between the cavity wall and the heating element experience the temperature change. This reduces the energy requirement. Both ordinary wound cartridge heaters and ceramic heaters are used. The advantage of ceramic heaters is their very high heating power in a small area. Their disadvantage is brittleness and thus susceptibility to breakage.

Induction heating and water cooling

In induction heating, as in a cooktop, heat is generated by a coil through which alternating current flows (the inductor). An alternating magnetic field induces eddy currents in the material, generating heat directly in the material, whereby the depth of penetration depends on the frequency of the alternating current. The advantage of this technique is the very precise control over the layer to be heated, i.e. only very small areas of the mold pass through different temperatures. The disadvantage is the high technical effort, because the inductors have to be integrated into the mold and powerful generators for the alternating current are necessary as peripherals next to the production equipment.

3.5.2 Implementation

Conventionally drilled temperature control loops

Temperature control is usually achieved via a system of intersecting holes. Screw-in sealing plugs provide for desired changes of direction. Figure 3.48 shows the principle behind cooling an oval container. In the mold, the holes follow the route of the contour and are plugged laterally in each case. It is clear that the channels cannot have an optimum distance to the cavity wall. The core could be temperature-controlled similarly. Here, however, this becomes problematic because the plugs are part of the cavity surface. In all likelihood, they will cause gloss differences on the part surface because they are not made of the same material as the mold core and thus will have different cooling conditions. The cooling circuit for the core is routed through vertical riser holes via an underlying plane.

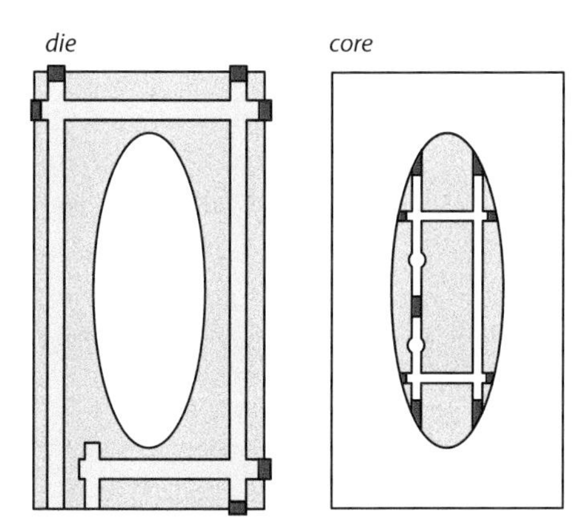

Figure 3.48
Cooling circuit via branched and plugged deep holes

Standards for cooling slim cores

The smaller the cores become, the more difficult it is to cool them via intersecting bores. Standards with which simple bores become a force-flooded temperature control channel are helpful here (Figure 3.49).

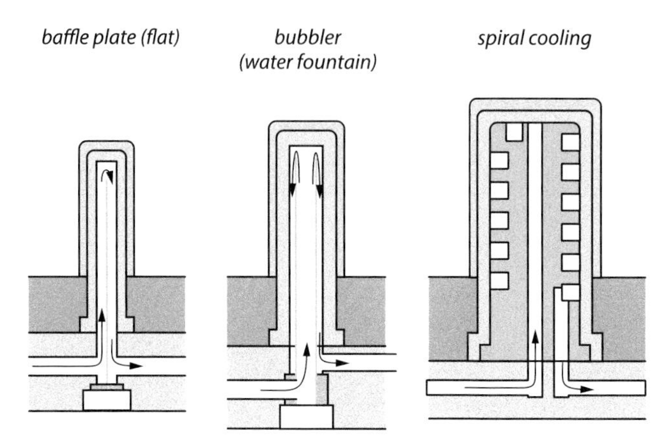

Figure 3.49 Core cooling by means of spiral cores or deflectors

Spiral cores are available in diameters from 12 to approx. 50 mm; thus cores up to approx. 60 mm diameter can be temperature controlled in this way. However, it should be clear that the flow cross-section represents a constriction that reduces the flow. The spiral cores inevitably have some play in the bore, so that part of the temperature control medium can also flow over the flanks of the spiral.

For cores smaller than 12 mm in diameter, finger cooling systems are suitable, in which the temperature control medium enters the tip of the core through a riser bore and flows back along the outside. Alternatively, cooling plates can be used which guide the temperature control medium through the core in both straight and coiled form. The remaining flow cross-section should always be borne in mind.

For difficult contours, a core can be manufactured using special processes, through:

- Vacuum brazing
- Diffusion welding
- Laser sintering

Multi-part cooling channel assemblies

In vacuum brazing, a core is brazed together from several individual parts. For example, the desired cooling channels can be milled into a surface and soldered together with the contour-forming counterpart (Figure 3.50). The comparison with conventionally drilled temperature-control channels shows the following:

- The layout of the channels can be designed much more freely; the core elements to be connected do not necessarily have to be planar (Figure 3.51).
- The channel cross-section does not have to be round; a rectangular contour has a larger surface area and thus a more favorable shape for heat exchange. However, consideration should also be given to the fact that flow in a corner area is somewhat slower than in the middle of a flow channel and thus the heat transfer or the efficiency is reduced.

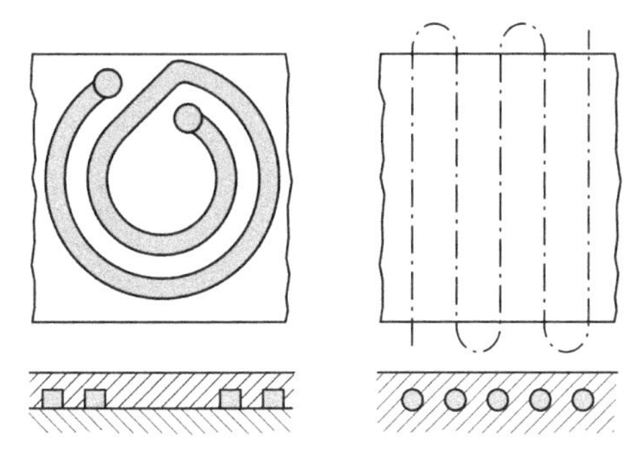

Figure 3.50 Conventional drilled cooling channels (right) compared to a milled multi-part cooling channel design (left)

A disadvantage of soldering can be separation of the soldered individual parts if this is in the area of the contour-forming cavity. The solder is also on the surface in small proportions, at least at the cut edge, and will behave differently from the base material during final texturing with an etching process, so that surface defects may also occur here. However, it is possible to insert copper pins in addition to the milled cooling channels to facilitate heat dissipation from areas that are difficult to access. In this way, very slim cores or mold slides can be temperature-controlled (Figure 3.51).

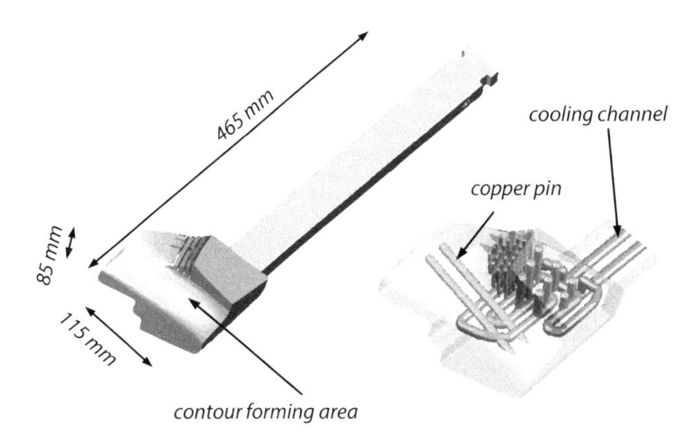

Figure 3.51 Vacuum-brazed core [image source: Contura, Menden]

In diffusion welding, core contours consisting of several elements are welded together in a process that is comparable to vacuum brazing. This welding takes place in a vacuum at temperatures just below the solidus temperature of the mold steel without any additional material. This results in a very good bond that is barely distinguishable from the properties of the base material. The final machining of the contour-forming mold areas takes place after the welding process.

Laser sintering (3D printing)

In laser sintering, a mold core is "3D printed". In this process, a layer of metal powder is spot-melted and welded with a laser beam (Figure 3.52). All areas that the laser does not hit remain powdery. Build-up of the mold core takes place layer by layer. The non-fused powder serves as support material for the solid areas that are formed above it. Finally, the powder can be shaken out and blown out. The disadvantage of this process is the high cost of approx. ca. 8 €/cm³. It is possible to print additional shapes on existing milled contours, thus keeping costs low.

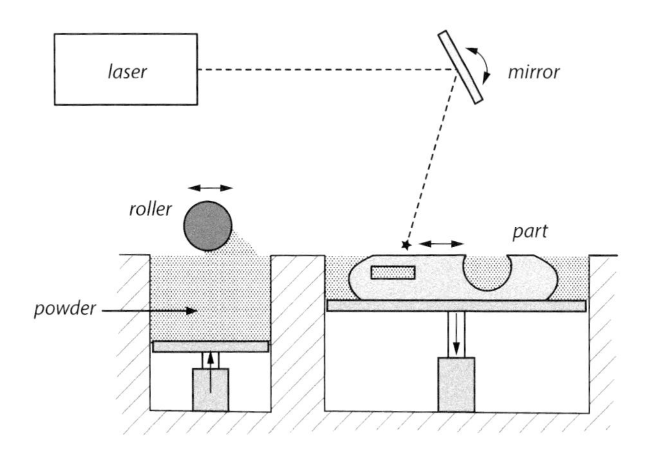

Figure 3.52 Generative build-up of cores by laser sintering

Design freedom is very high with laser sintering (Figure 3.53), but care should be taken to ensure that the the temperature control medium flows through all temperature control channel areas. The smaller the cross-sections to be flooded, the higher is the pressure requirement and, in addition, the likelihood of turbulent flow with good heat transfer behavior is reduced.

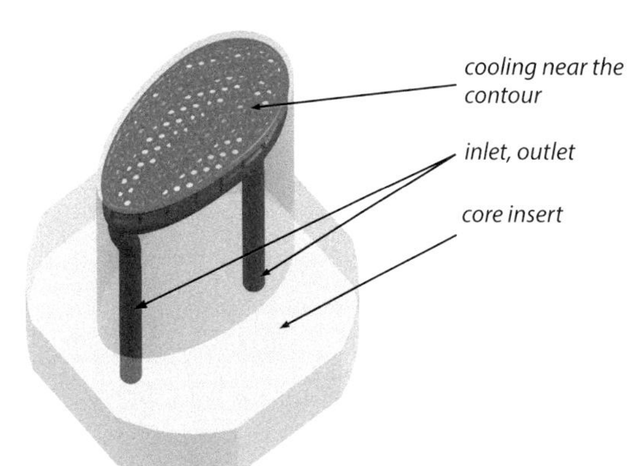

Figure 3.53
Mold core with generatively produced near-contour cooling [image source: Werkzeugbau Siegfried Hofmann GmbH, Lichtenfels]

■ 3.6 Demolding

Component must remain on the ejector side during demolding

The design must ensure that the component remains on the ejector side of the mold when the mold is opened. This can be achieved, for example, by selecting slightly smaller draft angles on the ejector side.

In addition to a suitable component design with the largest possible draft angles, reliable demolding requires a moving mechanism to be integrated into the mold. A fundamental problem here is the availability of installation space, and also cooling.

Usually, the “B side” (the moving mold half) is the ejector side, i.e. the ejector mechanism is housed within the mold. On the moving side, an ejector (bumper) is provided, usually a rod which can move parallel with the axis of movement of the mold and which activates the ejector system of the ejector side.

It is very important for the designer to consider that, when the machine is opened, the part initially remains on the moving mold side, otherwise it would remain on the “A side” and cause a process interruption, where the part would have to be demolded by hand.

Use of robots and handling systems

In principle, there are also handling systems and robots for removing parts from the mold space, but these cannot ensure immediate demolding. Because plastics become noticeably smaller during cooling, they shrink onto cores. For large parts of a certain part height, release from the mold core requires a movement perpendicular to the cavity without tilting or canting.

The first millimeters of demolding require quite high forces, depending on the demolding slope, with the ejectors pushing through the part if the demolding system is poorly designed, triggering a process interruption. This also happens if the selected cooling time is too short and the plastic is not yet stiff enough for demolding.

Ejector protection

For safe functioning of an injection mold, it must always be ensured that the demolding mechanism is in its zero position before mold closing; the ejectors must be retracted. Otherwise, they could hit the cavity surface of the nozzle side and damage it. This safeguarding can be done by:

- Monitoring of limit switches on the ejector plate
- Mechanical coupling with the ejector on the machine side, which is itself secured by the machine
- A combination of spring actuation and push-back pins, where a spring pushes the ejector pin to the rear, end position and additional, usually large ejectors are positioned adjacent to the cavity and hit the nozzle side. The ejector pins are shorter than the push-back pins by the component thickness and thus cannot hit the cavity surface of the nozzle side (see Figure 3.17).

There are different concepts for demolding the injection molded part that vary with the part design. The designer should be aware of:

- The effort required for demolding from the mold depends critically on the shape of the component.
- Demolding is performed by a mechanism integrated in the mold and requires installation space that may be lacking for the installation of an effective cooling system.

The following classification should cover the usual questions.

3.6.1 Straight-Line Demolding in the Axial Direction of the Opening Movement

Demolding in one axis

Simple parts without undercuts can be produced with a standard mold (two-plate mold) (Figure 3.54). The ejector system consists of several pins which press on the surface of the component. For a demolding stroke to completely demold the component, there must be free space for the ejector plate to move inside the mold (ejector box). As a result, the cavity may be poorly supported in the direction of the ejector system. It is of course possible to use support pillars through the ejector plate so that the cavity plate is supported via the mold base plate.

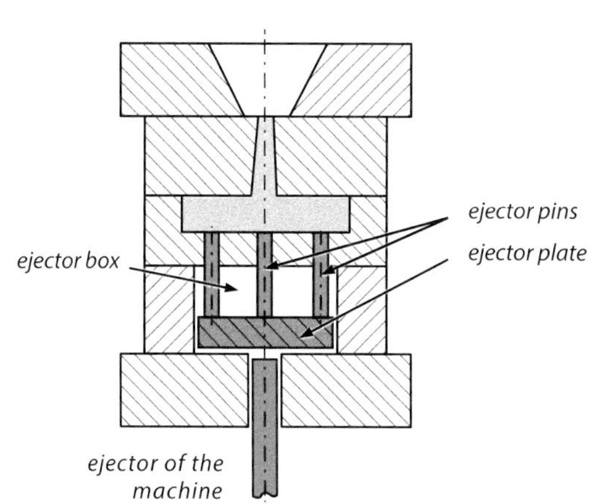

Figure 3.54
Two-plate mold with ejector system

Ejector pins, special features during installation

Ejector pins are guided with close tolerances in the front area (Figure 3.55) to prevent melt from penetrating into the gap between the cavity plate and the ejector. In the rear area and also in the ejector retainer plate, the ejectors must be provided with sufficient clearance to facilitate easy movement easily, even if the mold heats up unevenly.

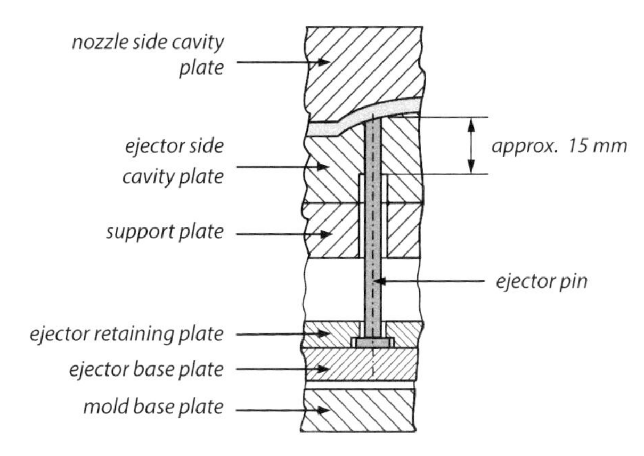

Figure 3.55 Ejector pin with guide

Number of ejectors

The number and position of ejector pins is difficult to specify, because the forces acting on a single ejector are difficult to calculate. It is important that the component be released from the mold at all points at the same time, because otherwise it will tilt in the mold contour and cannot be demolded at all. The size (diameter) of the ejector pins is also selected largely on the basis of experience. Ejectors that are too small may pierce the component instead of pushing it off the core, especially at high demolding temperatures. Ejector pins are available in round and rectangular cross-sections. The rectangular, flat shape (flat ejector) is often selected when narrow ribs have to be demolded. If necessary, the rib should be thickened locally to increase the pressure area and round ejectors can be used because a rectangular recess in the mold plate is more complex and expensive than is simple drilling for a round ejector.

Anti-rotation protection

Ejector pins are through-hardened and ground to a close tolerance. If the contact surface of the ejector is not parallel with the surface of the ejector base plate, the pin must be adapted to the contour of the cavity and it must be secured against rotation. Anti-rotation is possible, for example, by grinding the rear thickening of the pin flat on one side and machining a corresponding counter-contour into the ejector retaining plate.

Ejector sleeves

Ejector sleeves are used in combination with fixed contour pins for demolding screw bosses. Contour pins are not hardened and are used for the impression of blind bores (Figure 3.56). Contour pins and ejector sleeves have very tight tolerances and are available as unit standards. The sleeve pushes the wall of the boss away from the contour pin, which means that longer screw-in holes can also be realized without the risk of the component sitting too tightly in the ejector side of the mold in the area of the screw-in hole and not being able to be demolded well.

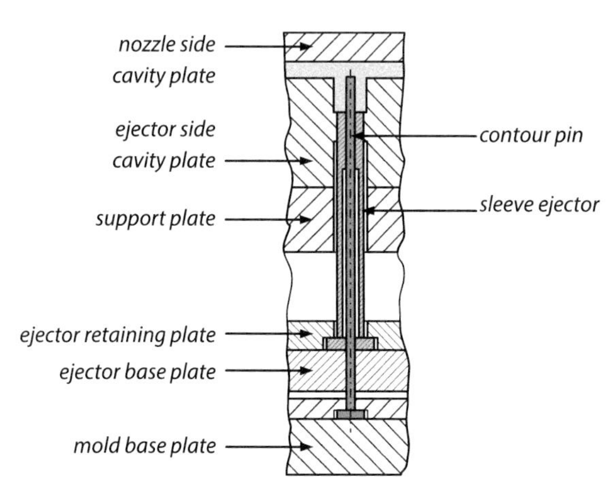

Figure 3.56
Ejector sleeve in combination with contour pin

Stripper plate mold

Easily demoldable, largely flat components with large demolding angles and no ribs or areas onto which plastic can shrink can be demolded with a stripper mold (Figure 3.57). Demolding here takes place via a stripper plate that presses on the edge of the component and not via ejector pins that hit various surface areas of the mold. Compared to the normal mold, this mold concept has higher rigidity, and the component is better supported in the direction of the ejector side.

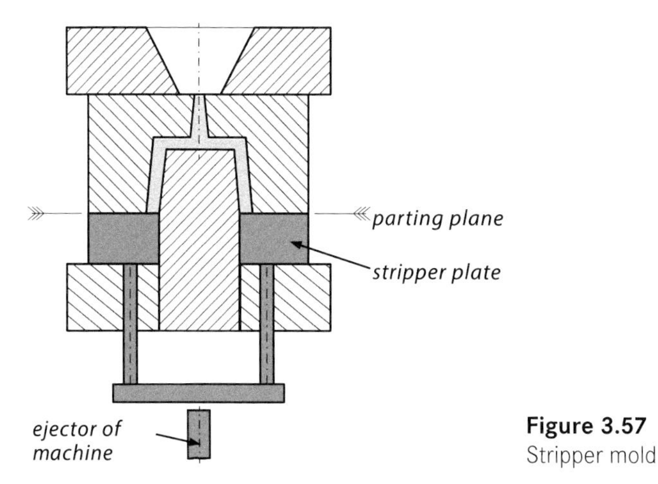

Figure 3.57
Stripper mold

Double ejectors allow two movements

For a more complicated part design, a multi-stage demolding process may be necessary. The part in Figure 3.58 shows the edge of a lid which has an undercut at the edge. If the plastic used has good ductility, the part can be pushed off the core via a forced demolding operation. However, this is only possible if the outer edge is exposed, because the surrounding mold contour would prevent elastic deformation of the plastic. In this case, a two-step ejector can generate two axial movements in the mold with only a single uniform stroke of the machine ejector. In the first step, the part is lifted with the core, exposing the outer edge. In the second step, the core stops and the ejectors continue to move, forcing demolding. Two-stage ejectors are available as standard parts.

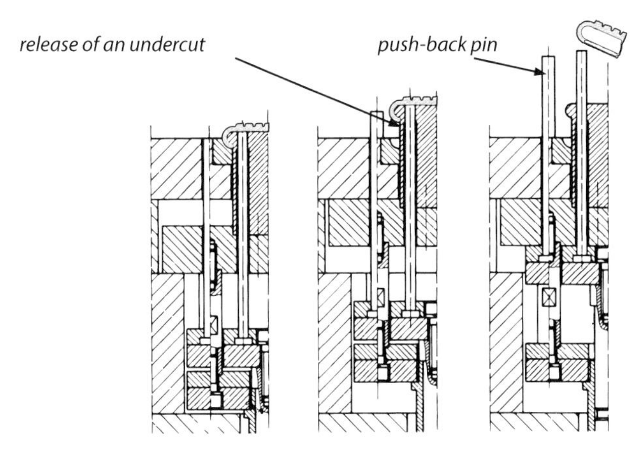

Figure 3.58
Two-stage ejector

Latch pulls control the movement of multiple plates

A two-stage movement also enables a latch lock, which is usually used for the controlled movement of two parting lines. This application becomes necessary if, for example, a component without a hot runner is to be injected not only at the edge but also at any points on the component surface. The cold runner system can then be demolded from a second parting line.

The pawl puller in Figure 3.59 opens first parting line B and then parting line A in succession by lifting pawl 3 after a first opening stroke via the gate on counterflap 6. Latch locks are available in different designs as standard parts.

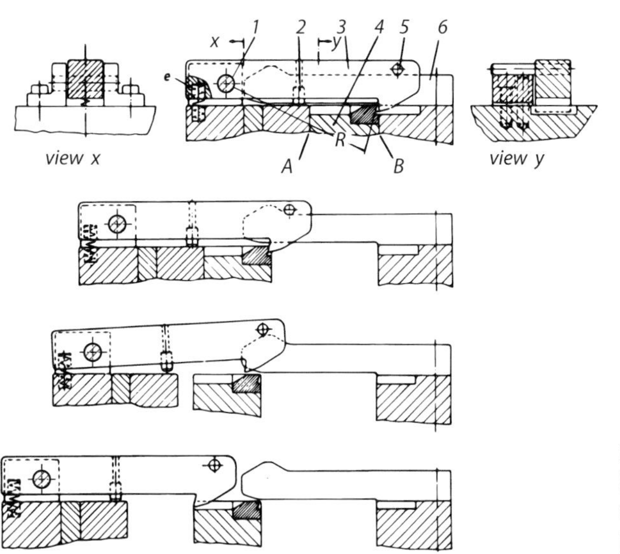

Figure 3.59
Latch lock for controlling the opening of two parting lines [image source: Mörwald]

Air ejector

Components that are very easy to demold can also be demolded pneumatically, largely without mechanisms. Figure 3.60 shows the demolding of a lemon squeezer using an air ejector. A spring-actuated plunger is advanced by compressed air, with the stroke being less important than the air blow. In this case, the spring's only function is to press the plunger securely into the retracted and closed position for the injection phase. Air ejectors are also frequently used in combination with the demolding options already mentioned, especially for relatively tall and steep-walled containers. During rapid demolding, insufficient air can flow between the core and the component, so that a certain negative pressure builds up. With the air ejectors mentioned above, this problem is easily solved.

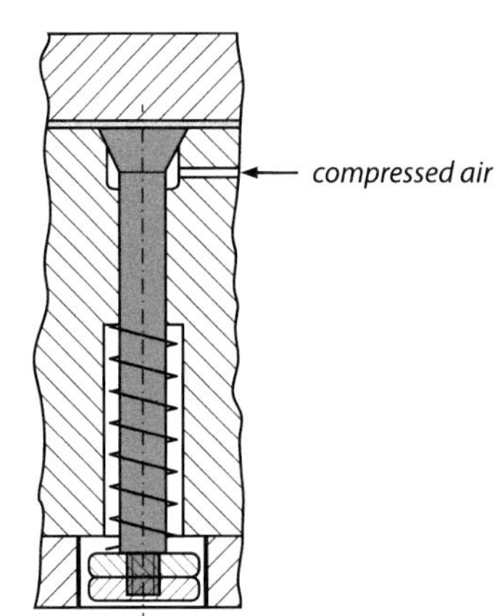

Figure 3.60
Air ejector [image source: Mörwald]

3.6.2 Demolding of Contour Areas That Are Not Parallel with the Opening Movement

Angled bolts guide slider in lateral movement

A complicated component design may also have undercuts, i.e. areas that cannot be demolded directly. These component contours have to be formed via separate mold elements, whereby these elements are moved transversely to the opening direction of the mold. A standard element is a slider, which is guided over an inclined pin during mold opening and is thus displaced laterally (Figure 3.61).

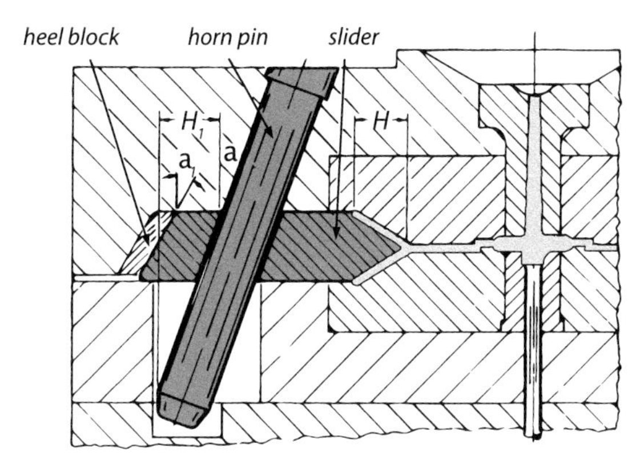

Figure 3.61
Sliders can be moved sideways by means of horn pins [image source: Mörwald]

The following must be observed during execution:

- The slider itself must be well supported in the mold, because the injection pressure acts on the entire contour-forming surface, and quite high lateral forces can result. Support is provided by the closing wedge, i. e. heel block.
- The horn pin (angle pin, slide pin) must not make contact with the slide under clamping force. The bolts are hardened and could break under transverse load. Therefore, the slide bore is oversized relative to the pin. In Figure 3.61, the bolt will be in contact with the right face during opening and will shift laterally. During closing, the contact shifts to the left side of the bolt. The angle of the closing wedge is slightly larger than that of the inclined bolt. When closing, the heel block (lock block) takes over the contact and pushes the slider slightly away from the horn pin.

A slide core requires space in the mold and, depending on the depth of the undercut, an opening stroke. It may be necessary to realize the movement of a cross slide via a core pull. Core pulls are contour areas that can be moved independently of the mold opening movement; this is usually done with cylinders that can be moved hydraulically (Figure 3.62).

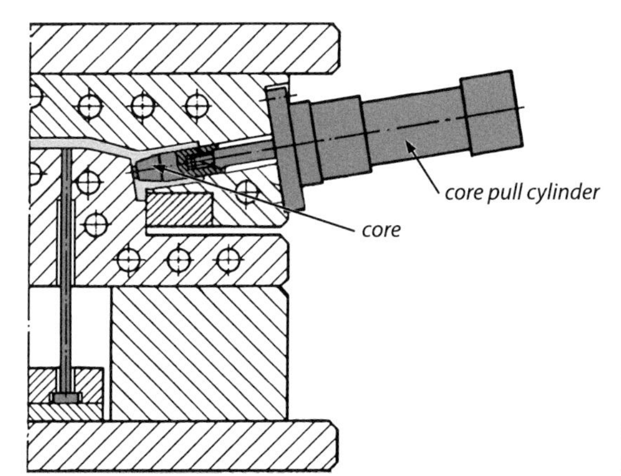

Figure 3.62
Core pull for lateral movement of a core [image source: Mennig]

In the case of core pulls, the following must be taken into account:

- Locking cylinders must be used, because even here the injection pressure would easily push the cylinder back again.
- The end positions of the core pulls must be secured with limit switches, because closing the mold can easily damage a core that is not in the correct position.
- In the case of multiple core pulls, appropriate control on the part of the machine is necessary, because the core pulls often have to be actuated in a certain sequence.
- Any type of core pull or slide must be able to withstand sufficient temperature. Especially with slim cores, removal of the heat of the molten plastic can be problematic.

With this in mind, due consideration should be given during the design process to whether it might be possible to change the shape of the part, thereby simplifying the mold design and eliminating the need for transverse movements.

3.6.3 Demolding of Internal Undercuts

Demolding of internal undercuts

In principle, slides can also be used for internal undercuts, by which is meant contour areas inside a component that are formed over the core. The smaller the components become, the more difficult this becomes, because a slider must be supported by a closing wedge and it must have room to move inward (Figure 3.63). Here, too, it should be borne in mind that the space required for the mechanism restricts the effective build-up of a temperature control system for the core.

Special ejector for demolding small undercuts

For small undercuts, standardized solutions are often available in the form of standard parts. Figure 3.64 shows an example of an ejector with a folding mechanism. The upper part is pressed to the right by a leaf spring not shown here when the ejector is extended. In this system, the injection pressure is supported by a shoulder which rests on a support edge in the retracted position.

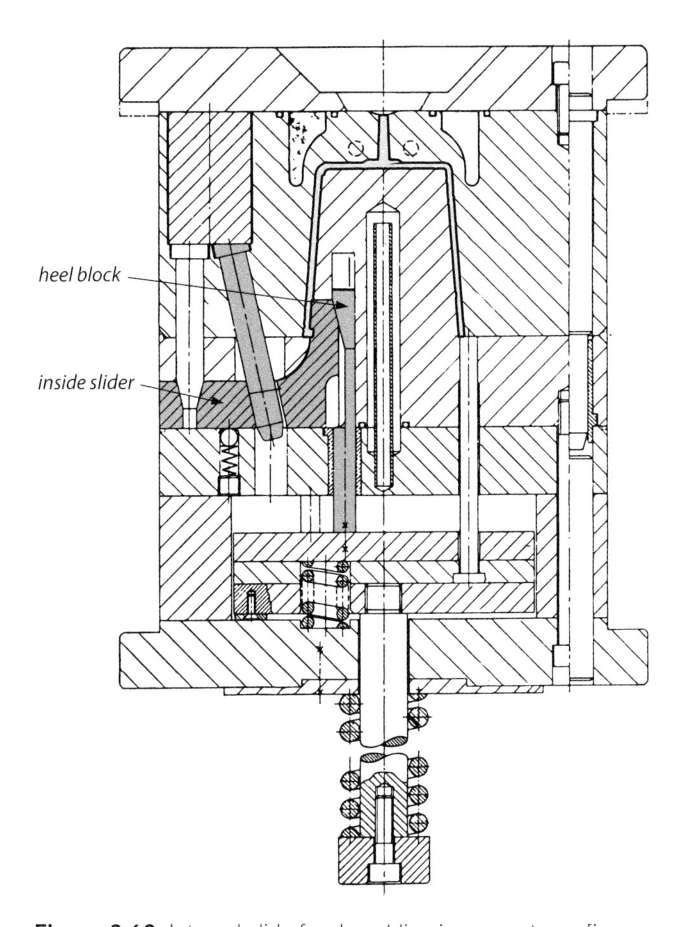

Figure 3.63 Internal slide for demolding inner contours [image source: Gastrow]

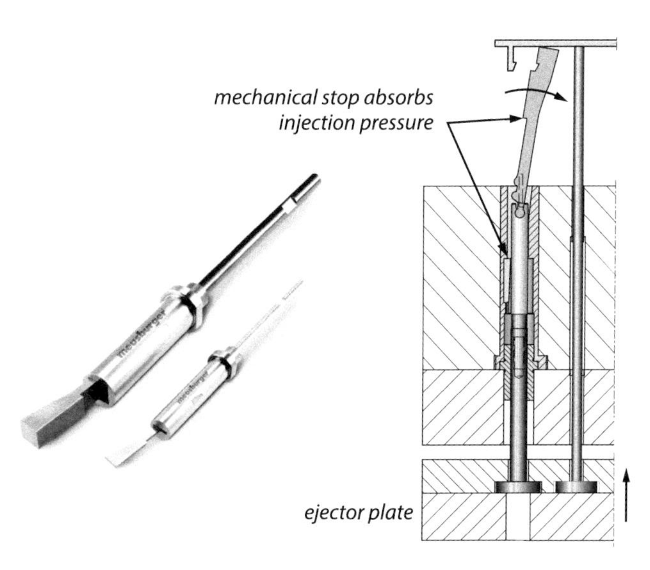

Figure 3.64
Ejector with movable spring joint for clearance of undercuts [image source: Meusburger]

Collapsible cores

Larger internal undercuts, including circumferential ones, can also be demolded with collapsible cores. In this case, the core consists of a large number of segments, each of which is guided by a control core. Figure 3.65 shows a collapsible core for demolding a thread. The narrower segments, marked in dark gray, are guided in a T-slot at a shallower angle than the wider segments. As a result, the narrower segments move inward more quickly when the control core moves downward, making room for the wider segments.

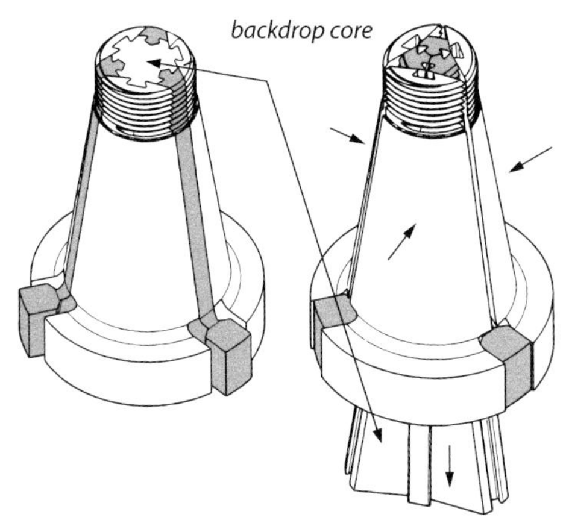

Figure 3.65
Collapsible core whose segments can be moved inward by moving the control core

Collapsible cores allow very rapid movements, because the control core can already be moved with the opening movement of the mold. With large undercuts, however, a long travel for the control core will be necessary, i.e. the mold will have a large installation height. Here, too, it must be borne in mind that the segments have to be sufficiently tempered.

3.6.4 Demolding of Internal Threads

Unscrewing molds

Internal threads can be demolded with drop cores or they must be unscrewed before demolding (Figure 3.66). The rotary drive can be provided by separate drive motors or by steep-thread spindles or toothed racks, whereby the above-mentioned mechanical solutions can also be coupled with the opening movement of the machine.

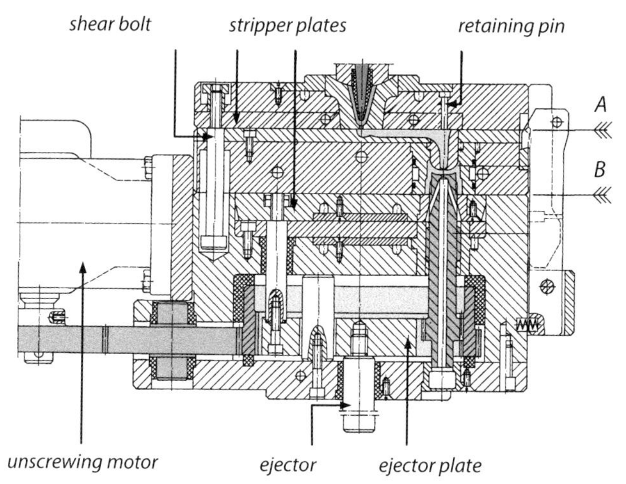

Figure 3.66
Unscrewing mold
(4-fold sealing cap)

■ 3.7 Increasing Efficiency with Two Parting Planes

Applications with additional parting lines

Section 3.4.1.2 shows a mold design with an additional parting line. Here, the purpose is to use a cold runner to set the sprue position at any location on the part surface or to set additional sprue locations. This section shows how efficiency can be increased with an additional parting line. There are two different techniques here, each with its specific advantages:

- **Stack molds** have two identical parting lines which are filled and opened simultaneously. Thus, twice as many parts can be produced with the same mold size in terms of clamping area as with a standard mold of the same size.

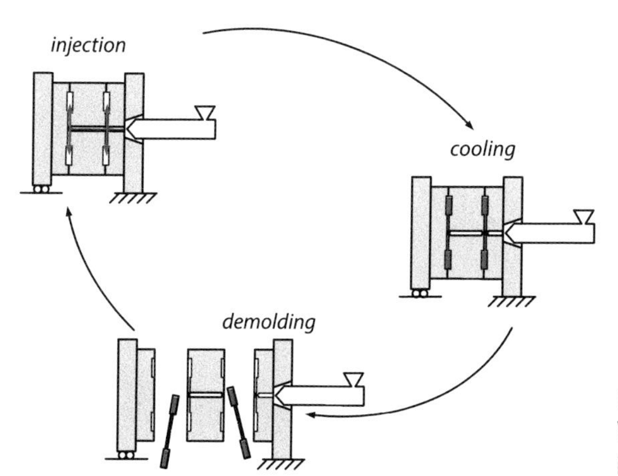

Figure 3.67
When a stack mold is opened, two parting lines open at the same time

- **Tandem molds** have two parting lines which are filled and opened at different times. It is possible and advantageous if the respective parting lines contain different components of an assembly. This is particularly advantageous for family applications.

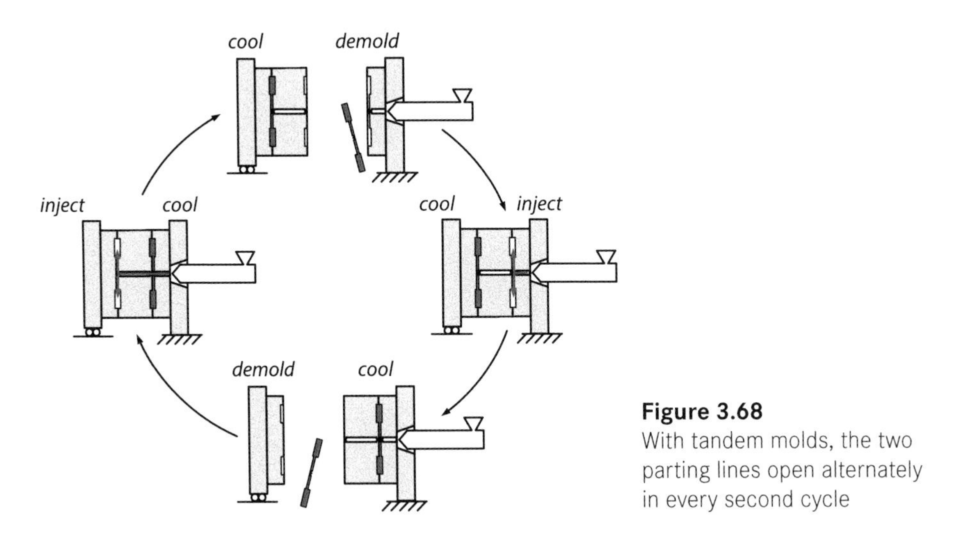

Figure 3.68
With tandem molds, the two parting lines open alternately in every second cycle

3.7.1 Stack Molds

Clamping force with opposite cavities does not increase

The clamping force of an injection molding application results from the pressure exerted by the internal pressure and the projected component area. If two components are located one behind the other in the direction of the clamping movement, only the mathematically required clamping force of one component is necessary, so that the machine can remain closed under the high process pressures.

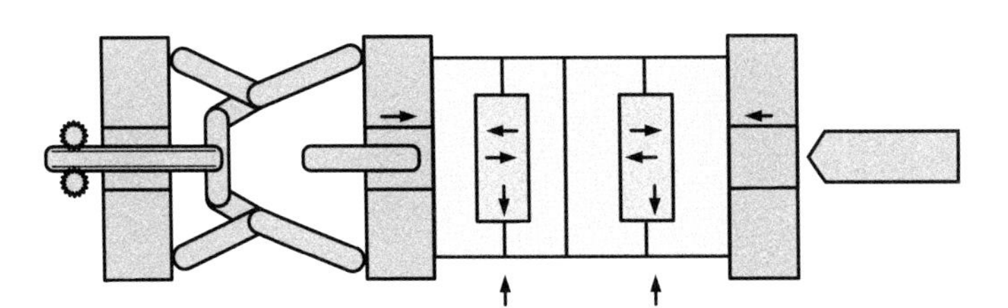

Figure 3.69 The forces in opposite cavities cancel each other out

Pressure always acts in all directions. If two components exactly overlap in their area, the pressures of the two components, each pointing in the direction of the other component, will cancel each other out. Therefore, only the outward pressure needs to be taken into account for the clamping force.

In this way, twice the production quantity is possible with a machine of a given clamping force. Thus, if a machine can produce 4 lids for a package with a clamping force of, say, 1000 kN, a stack mold will be able to produce 8 lids with the same clamping force.

This increase in production output is possible in the same cycle time if the machine meets the following characteristics:

Increase in production output depends on machine equipment

- The plasticizing capacity must be so high that the melt required per cycle can be provided within the cooling time and, if necessary, the subsequent mold movement and component demolding.
- The injection performance must be so high that the melt volume can be injected in the same injection time as in a comparable standard mold. Specifically, this means that the one injection speed must be twice as high. It should also be borne in mind that the injection volume flow into the individual cavity is identical with that of the standard cycle, but the volume flow in the melt feed, i.e. in the hot runner, has to be twice as high.
- The opening width of the machine must be large enough to allow both parting lines to be opened when the mold is opened and to allow problem-free demolding. Stack molds are used almost exclusively for flat applications in the packaging sector. The opening stroke increases with increase in component height. For demolding, a stroke greater than twice the component height is required in any case. For this reason, stack molds are advantageous for flat components.

3.7.2 Tandem Molds

To understand tandem molds, it is important to know that the clamping force of the machine needs to be high only for a brief moment. During injection, the average cavity pressure increases with the amount of melt there in the mold. In the case of components that do not have very thick walls, the melt cools during the holding pressure phase, so that the pressure can only act close to the gate and finally no longer reaches the cavity from the driving screw. Calculations performed with simulation programs have shown that the clamping force is only necessary for about half of the cycle time, and only during injection and holding pressure. During the cooling phase, the mold can also remain closed without clamping force and absorb the heat of the plastic melt.

No clamping force is required during the cooling phase

A simple locking system can be used to alternately lock one of two parting lines, so that the respective unlocked parting line can be opened. Like a pawl-pull mechanism, a locking slide moves between two end positions after the mold is closed and presses two rollers into an undercut. In this way, one level is unlocked, and the second level is locked.

Locking system of tandem molds does not have to withstand clamping forces

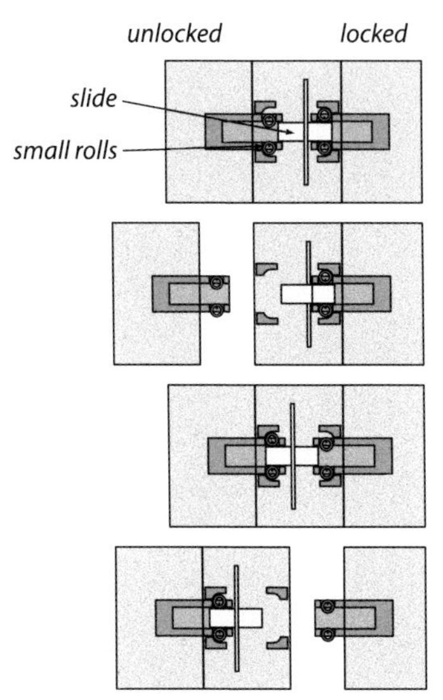

Figure 3.70
Locking system of a tandem mold

Melt is always injected into cavities of the non-interlocked parting line

The melt can be injected into the non-locked plane; this is important because, in the event of a malfunction and consequent over-injection of the cavity, the mold can breathe freely and the locking system will not be damaged. After the injection and holding pressure phase in this plane, the machine can open the mold. Because the level with the not yet cooled melt is locked, only the second unlocked level opens and the components here can be demolded.

Applications for tandem molds

Possible applications of tandem molds are:

Family molds contain undersized elements of an assembly

- **Family applications** – There are often several plastic parts that belong to one assembly, e. g. a housing consisting of an upper and a lower part. In such component families, the various individual elements are usually different, and each requires its own process settings. Family applications are initially unpopular from a production point of view because independent process settings are not possible with a standard mold. In the case of tandem molds, the individual parts can each be molded one after the other in cyclical alternation, each with its own process settings. There are several advantages here:
  - Parts of an assembly always come out of the machine directly one after the other and can either be joined directly or at least transported or stored together to fit. Without a tandem mold, the components would be molded on independent machines at different locations; often this is also done on one machine with several molds, staggered in time.
  - When plastics are colored with liquid colors or masterbatches during the plasticizing process, minor color deviations cannot be ruled out. When such slightly different-colored components of an assembly are joined together, this is often

enough a reason for complaint. With tandem molds, such color differences can be ruled out, because a difference is not visible from cycle to cycle.

- **Right/left-hand applications** – Mirror-symmetrical contours are often used, especially for automotive components. Because both parts are always used for an automobile, two molds and two machines are needed for production, or a tandem mold is used on one injection molding machine. In the case of spare parts production, it is possible to use a tandem mold with only one of the two levels and leave the second level permanently locked.

Mirror-symmetrical applications

- **Thick-walled components** – The thicker the components become, the longer the necessary cooling time and thus one cycle. In standard applications, the sum of cooling time and machine movement time (open, demold, close) is about half as long as the total cycle time. In this case, the production output of the machine doubles when a tandem mold is used.

Components with long cooling times

Tandem molds can reduce production costs by 25% if only the necessary direct production costs are considered rather than the overheads. These include material costs (approx. 50%), labor costs (approx. 5%), machine costs (approx. 25%), and mold costs (approx. 20%). The savings result because only one machine is needed for the same quantity of parts. A tandem mold costs about as much as two standard molds, because the cost of the necessary locking is negligible compared to the machining costs for the two cavities.

Tandem molds can save the need for an injection molding machine

It is interesting to note that there is also a cost advantage for thin-walled components and that component quality tends to be increased. This is due to the additional cooling time. The machines can only open when the plasticizing time for the next shot in the second level is finished. This means that the first level remains closed for this time difference (plasticizing time minus cooling time), and the components inevitably arrive at demolding somewhat further cooled down and thus tend to be more dimensionally stable.

Supplemental cooling time for thin-walled components

Usually in production, efforts are made to shorten the cooling time and thus increase production efficiency. This often leads to increased effort and higher costs for effective cooling already during mold construction. When tandem molds are used, the additional cooling time leads to an increase in performance of 30% and more, even for thin-walled components in fast-running machines.

The speed of a tandem cycle can be estimated if the cycle time components for the essential functions are known. A distinction needs to be made here between machines with or without parallel movements, because hydraulically driven machines often cannot perform the mold movement (opening, demolding, closing) simultaneously with the rotary movement of the screw during plasticizing. For this reason, high-performance injection molding machines with a higher plasticizing capacity and parallel drives are often used.

Cycle time for tandem molds depends on the drive type of the injection molding machine

standard mold 4-cavity - 3000 parts/h

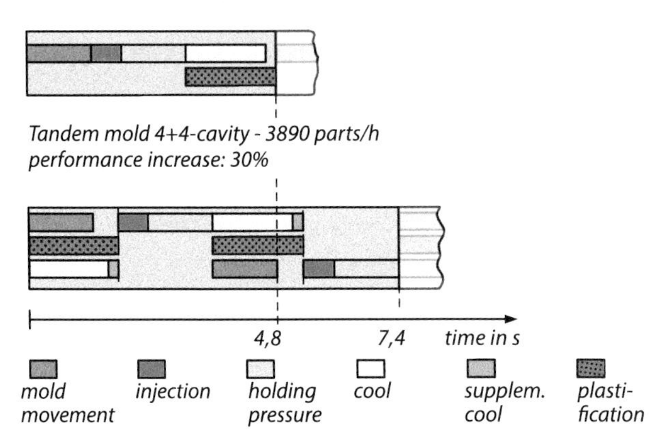

Figure 3.71 Simultaneous process components for tandem molds depending on the drive type of the injection molding machine

Calculation of the cycle time of tandem molds

machine without parallel movement

$$t_{cycle} = t_{inj.} + t_{hold\,press.} + max \begin{pmatrix} t_{cool} + t_{machine\,movem.} \\ t_{machine\,movem.} + t_{inj.} + t_{hold\,press.} + 2t_{plastif.} \end{pmatrix}$$

machine with parallel movement

$$t_{cycle} = t_{inj.} + t_{hold\,press.} + max \begin{pmatrix} t_{cool} + t_{machine\,movem.} \\ max \begin{pmatrix} t_{mach.\,movem.} \\ t_{plastif.} \end{pmatrix} + t_{inj.} + t_{hold\,press.} + 2t_{plastif.} \end{pmatrix}$$

Figure 3.72 Cycle-time calculation for tandem molds

Requirements on Machines for Use with Tandem and Stack Molds

Drive equipment needed for use with tandem molds

Machines for use with stack molds have to be designed for a very large capacity. This concerns:

Opening width of the clamping unit

- Clamping unit with increased opening width (distance between the fixed and moving mold fixing plate in the maximum possible opening position). If two parting lines are located one behind the other, the mold installation height automatically increases. Therefore, the machines for such molds usually have an enlarged opening width.

Plasticizing and injection performance

- Injection units with larger screws allow higher plasticizing and injection performance (melt per unit time). In the case of tandem molds, only the plasticizing capacity has to be greater, because, especially in the case of machines without a parallel drive, the machine cannot open until the metering process has been completed. In the case of stack molds, two parting lines are filled at the same time, which automatically means that twice the injection rate is required.

- The drive power must be large and parallel movements should be possible. Especially in the case of packaging applications, the wall thicknesses of the components are small and thus the necessary cooling times are small too. So that the plasticizing time does not slow down the entire cycle, such machines should allow several movements at the same time, which in this case means the rotary movement of the screw for plasticizing and movements of the clamping unit for opening and demolding.

Parallel movements for metering and tool movements

For the operation of tandem molds, the injection molding machines require additional equipment:

- It must be possible to alter the process parameters for injection speed, holding pressure level and holding pressure time in cyclical changes. That is the only way to get the best process settings needed for the quality of different-sized components in a component family.

Rapid changes of injection parameters

- The locking system of the tandem mold is controlled like a core pull via the operating unit of the machine. The process parameter set for the non-locked plane has to be active in each case.
- It must be possible to deactivate the control of the interlocking system. Then an operator can concentrate on optimizing the process setting of a component during the setup process. When the process parameters for the first level are set, it can be locked and the operator can optimize the process for the second level. If it becomes necessary to produce components of only one level, e. g. in the case of spare parts production, the other level can be completely shut down at any time.

Triggering of the interlocking system

- The opening width of the machine must be large, because the molds have considerably greater mold installation heights. Unlike stack molds, it is not just flat components that lend themselves to tandem molds. In principle, a tandem mold is twice as high as a standard mold. The opening stroke does not have to be increased, because only the stroke for demolding one level is ever required.

Large possible opening widths of the clamping unit

3.7.3 Design Features of Stack and Tandem Molds

Position of the ejectors with two parting lines

The ejector unit of molds that have two levels for the components is directed outward, because this allows the path of the melt to be symmetrical for both levels. This is particularly important for stack molds, because here both levels are filled at the same time. If the melt is first fed from the nozzle into the center plate, the paths from here into the first and second levels are of equal length. If the flow paths into the cavities were of different lengths, different process parameters would be necessary.

The ejectors on the nozzle side are actuated by separate short-stroke cylinders which, in the simplest case, are mounted on the side of the mold and move an ejector plate located in the nozzle side. It should be noted here that the melt feed runs through the center of this ejector unit. If the melt is fed via a hot runner, it must be borne in mind that here the heat of the hot runner may have to be isolated to the ejector.

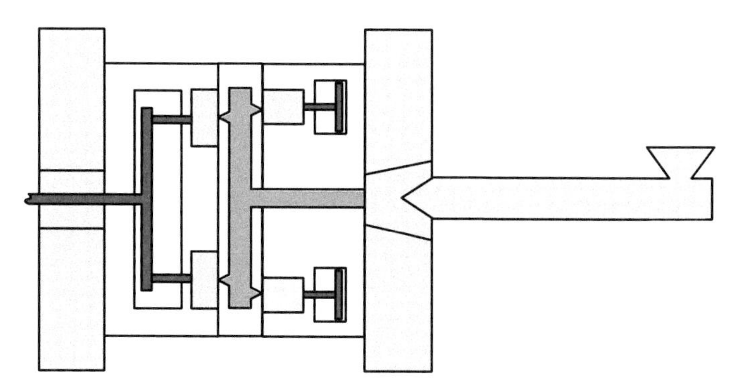

Figure 3.73 Position of the ejector sides for molds with two cavity levels

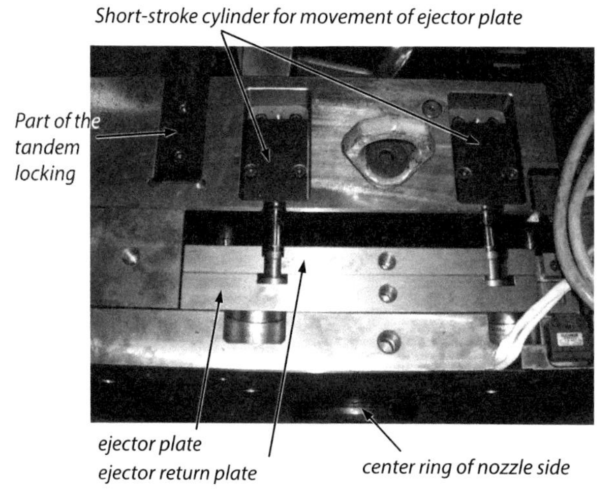

Figure 3.74 Drive of the ejector plate, with laterally mounted hydraulic short-stroke cylinders

Stack molds have some special features:

Support of the center plate

- The center plate must be supported because, in the mold-open position, it is initially without contact to the mold halves attached to the machine. This support is provided either by a sliding guide in a yoke that transfers the weight of the center plate to the tie bars of the clamping unit. Or the support is provided by a sliding guide on the machine bed below the tie bars.

- During the opening movement, the two parting lines open synchronously at the same speed. This requires a mechanism that moves the center plate at half the opening speed of the opening machine plate. This is possible with a rack and pinion drive mounted on the side of the mold or a lever mechanism.

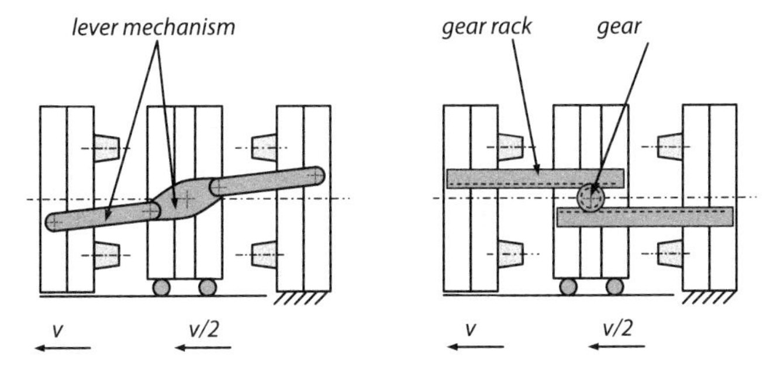

Figure 3.75 Mechanical synchronization of the opening movement of the center mold plate

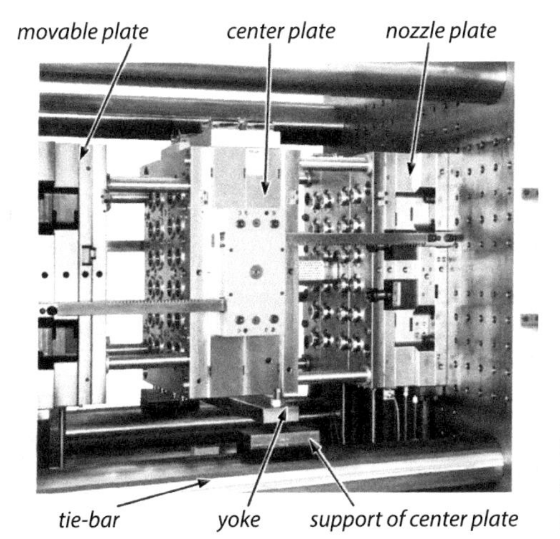

Figure 3.76 Support of the center mold plate on the tie bars

The special features of tandem molds are:

Support of the center plate for tandem molds

- The center plate does not have to be supported on the tie bars or the machine bed. Laterally mounted flat guides are advantageous, as they always support the locked center plate on one of the two outer-plate packages. This allows the tandem mold to open in the same way as an ordinary two-plate mold.

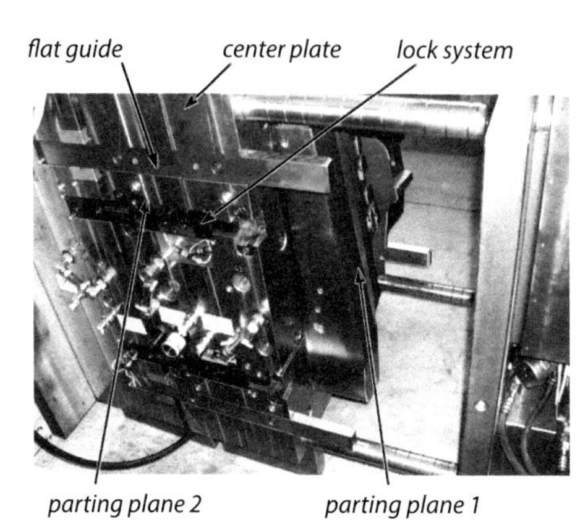

Figure 3.77
Support of the center plate of a tandem mold via lateral flat guides

3.7.4 Hot Runner Technology for Stack and Tandem Molds

For molds with two parting lines, the melt is usually fed into the cavities via a hot runner in the center plate. Three variants are possible for this.

Hot runner snorkel with permanent connection to the manifold

- The hot runner is permanently connected to the nozzle via a snorkel. With each movement of the center plate, the contact to the machine nozzle loosens. This can cause a little melt to escape from the hot runner of the center plate in the direction of the nozzle and, over many cycles, build up enough material between the nozzle and the snorkel to cause a malfunction. This version is therefore less suitable for very-low-viscosity melts.

Hot runner nozzle with separation to the hot runner manifold

- The connection between the hot runner in the center plate and a heated long nozzle in the nozzle plate is separable. The contact is loosened with each movement of the center plate. Here, it is possible to seal the contact point via a valve gate. Leakage from the heated long nozzle can be prevented by retracting the screw a little before the plasticizing process, thus relieving the hot runner. This variant has the advantage that molded parts from the parting line near the nozzle do not meet a hot runner snorkel during demolding and possibly be damaged.
- The melt can be deflected to the side in the nozzle plate. This may be necessary if the components are so large that direct passage from the nozzle to the hot runner in the center plate is not possible. It is also possible to seal the contact point of the two hot runners of the nozzle side and the center plate with a sealing system. The disadvantage here is the overall greater length of the hot runner.

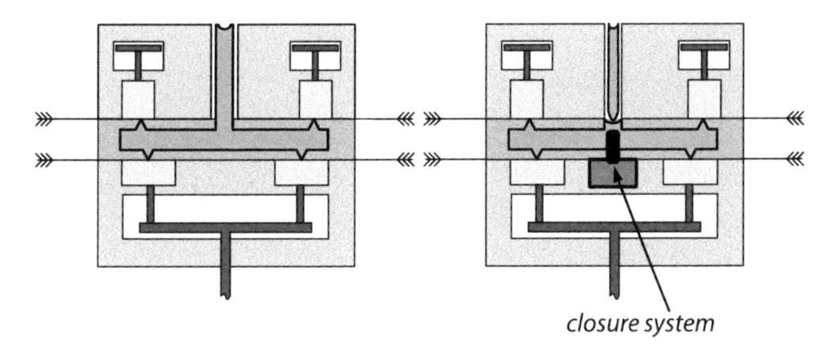

Figure 3.78 Hot runner variants for tandem molds; left: undivided, right: divided – the connection to the machine nozzle (sprue snorkel) is fixed

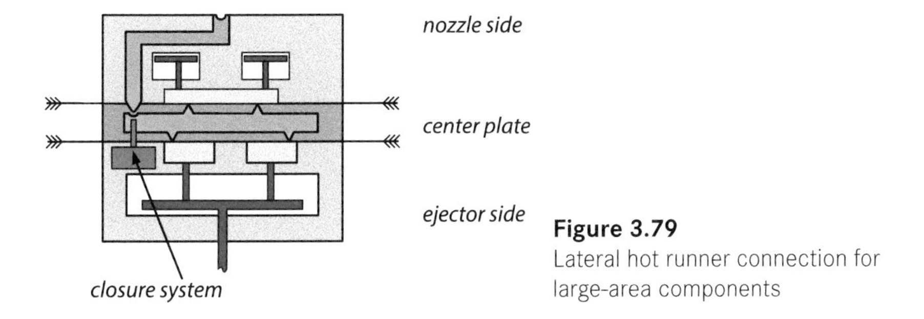

Figure 3.79
Lateral hot runner connection for large-area components

The separation of two hot runners in one mold is problematic, especially for stack molds, because melt can escape during the non-contact time, making direct contact and thus sealing of the nozzles to each other difficult. Here it must be considered that the tightness must be achieved solely with a metallic contact of the nozzles to each other.

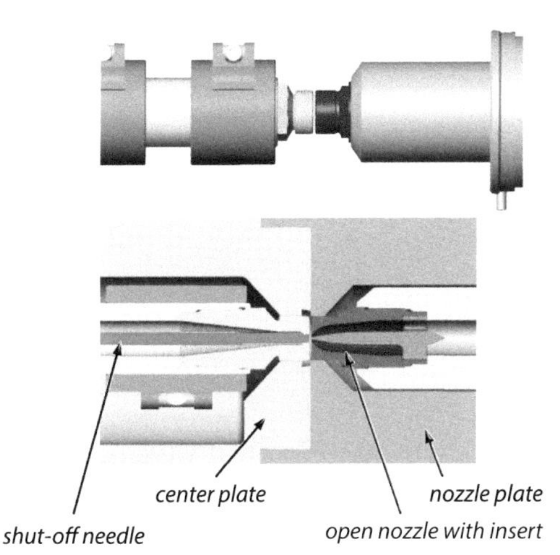

Figure 3.80
Sealing of the hot runner separation via metallic contact of the nozzles

Sealing of the hot runner separation in tandem molds

In the case of tandem molds, direct contact between the nozzles can be made via a plastic closure plate, which also acts as a cold sub-distributor. When installed, the nozzles are spaced apart and the gap between them is effectively a cavity. This closure plate is ejected in every second cycle with the demolding of the molded parts of the parting line on the nozzle side. In the subsequent cycle, this part remains in the closed nozzle-side parting line and the parting line behind it is opened. During the subsequent injection process, this cooling closure plate is broken through by the flowing melt.

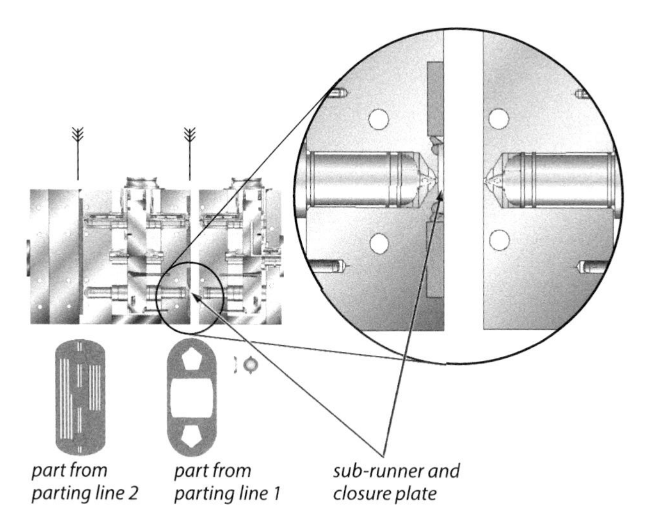

Figure 3.81 Sealing of the melt transfer point via an additionally injection-molded closure plate

Push-through molding for large-area components instead of hot runner deflection

The incoming melt builds up a very high pressure of over 1,000 bar during injection. This energy is sufficient to penetrate wall thicknesses of over 2 mm with a nozzle bore of approx. 1 mm. The already cooled material of the closure plate of these dimensions is so small that it melts completely on its way through the hot runner in the center plate and can enter the cavities without any loss of quality.

In tandem applications, the redirection of the hot runner around the component can be omitted for large-area components if the technique of the closure plate is transferred to the component in the front plane. Here, the plastic is in fact cooling while the cavity behind it is being filled. However, the small melt volume at the point of the breakout is so small that it can be demolded with the front cavity component even at high temperatures. Part quality is not affected by this.

4 Simulation

The lead time for developing components and assemblies is shortening all the time, partly because product life cycles are becoming shorter, and product variety is increasing. Simulations can save time because they enable the effect of measures taken to be predicted.

Simulation shortens development time

Frequently, different partners are involved in product development, each of whom has very different competencies and perspectives and is not necessarily in the same company (Figure 4.1). A customer requires a product which he/she may want to resell. He/she formulates his/her wishes and requirements in a tender specification. A supplier will turn the requirements specification into a specification sheet, which forms the contractual basis for the order. The statements in these specifications are binding; in concrete terms, the supplier will guarantee a defined product quality.

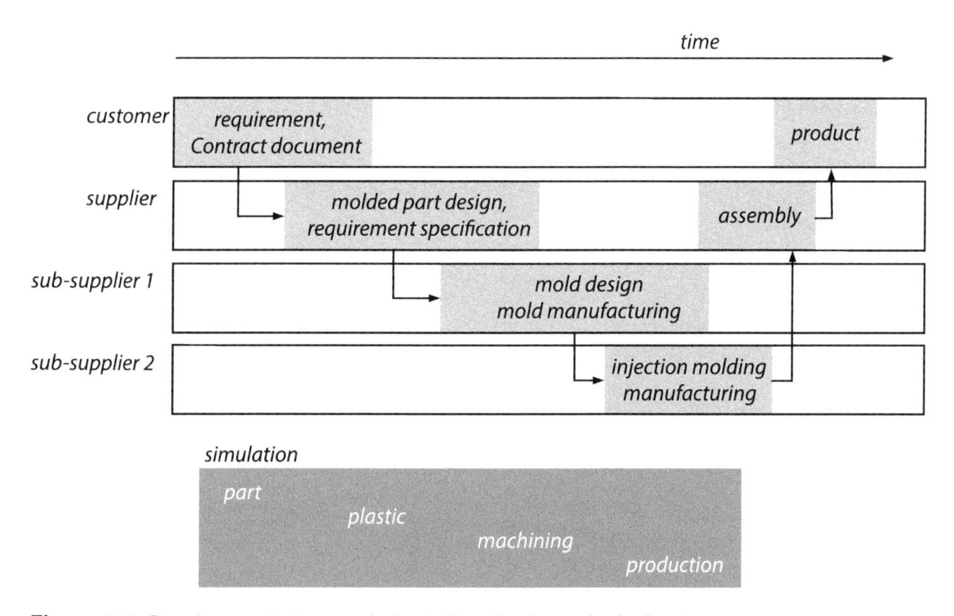

Figure 4.1 Development steps and simulations in chronological order

Different goals between customer and (sub)supplier

The supplier's design department may modify and design the product in detail. For example, while the customer has in mind the appearance and shape of a component, the supplier will also consider such aspects as manufacturability and possible quality degradation due to weld lines or component distortion. The same applies to the sub-suppliers who build the injection mold and produce the component themselves. The mold maker will specifically look at the feasibility of implementing the design with an injection mold.

Simulations are possible and useful to all parties. For the customer, it may be a question of how the component or assembly behaves under load; for the various suppliers, the focus may be more on injection molding simulation (rheological simulation) and the calculation of machining operations (CAM).

Difference between simulation and practice due to different process settings

The basic problem in the field of injection molding simulation is illustrated in Figure 4.2. For the simulation, production parameters (a_1, b_1, c_1, ...) have to be passed to the calculation program. The result of the calculation yields information, such as a high probability of warpage. The designer will make changes to the part design and, after some development loops, a tool will be ordered. In production, different setting values (a_2, b_2, c_2, ...) are certain to be chosen for the machine, because setting values can be estimated, but not predicted with accuracy. Surface defects will only be able to be optimized by a setter conducting trials during the start-up process.

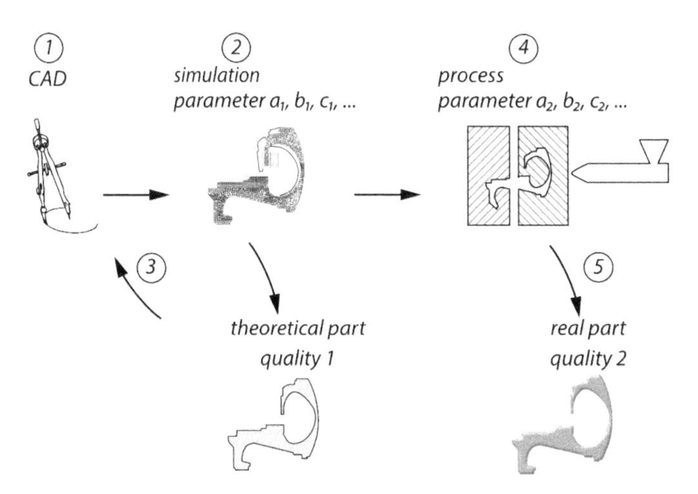

Figure 4.2 The results of simulation and production do not necessarily match up

Components should have the same quality regardless of process settings

In principle, the designer, together with the simulation expert, should design a component structure that always leads to the same quality, wholly independently of the production parameters (machine settings, e. g. temperature, loading speed, ...).

4.1 Goals of Simulation

Predictions about component quality

From the point of view of the customer or the supplier, the quality of a component comprises three characteristics. In the following, only the calculable characteristics are considered. If no theory has been formulated for the relationship between gloss and process setting, there is no formula for calculating gloss, so it cannot be calculated. The objectives for a simulation are therefore:

1. Good surface without deterioration due to weld lines or possible burns due to diesel effect.
2. Exact component dimensions without distortion.
3. Cycle times as short as possible, whereby the cooling process plays an important role in the simulation.

In the following, these goals are considered in a little more detail and the type of simulation required for each is explained.

4.1.1 Filling Simulation (Rheological Simulation) for Good Surfaces

Simulation of the injection process

The quality of a surface is not just determined by the gloss. It is also an advantage if no weld lines or flow lines are visible on the surface. To this end, it is very useful to simulate what happens to the melt during mold filling.

Hagen-Poiseuillle

The actual injection process is simple and can be simulated very reliably. According to Hagen-Poiseuille, the pressure requirement depends on the flow velocity v, melt viscosity η and the component shape, which is defined by the flow length L and the component thickness s.

$$\Delta p \sim \frac{Lv\eta}{s^2} \tag{4.1}$$

Melt flows faster in thicker molded part areas

Given different component thicknesses, the melt will have the same pressure requirement from the gating point in all flow directions but will flow into the thicker areas to a different length and at a different velocity (Figure 4.3).

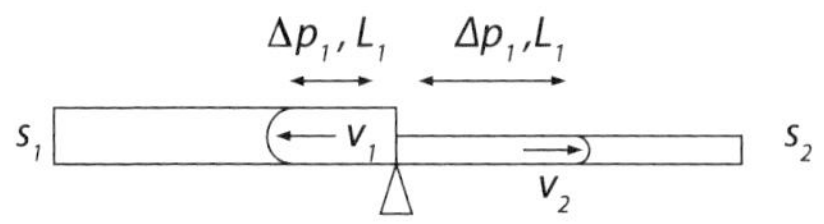

$\Delta p_1, L_1$ $\Delta p_1, L_1$

s_1 v_1 v_2 s_2

$$\Delta p_1 = \Delta p_2$$

$$\frac{K L_1 v_1 h}{s_1} = \frac{K L_2 v_2 h}{s_2} \quad \text{with } v = L/t$$

$$\frac{L_1^2}{s_1^2} = \frac{L_2^2}{s_2^2}$$

Figure 4.3
Physical fundamentals for filling simulation

The velocity *v* is the ratio of displacement *L* and time *t* and, because the flow front always reaches the different flow lengths at the same time in the areas of different thickness, the following very simple formula is obtained:

$$\frac{L_1}{s_1} = \frac{L_2}{s_2} \tag{4.2}$$

Graphical fill image determination

This can be used to determine the filling process graphically (Figure 4.4). From a gate, the melt swells to different extents according to the thickness at time 1. The two circular arcs can be connected in a simplified manner with a tangent. The same applies to other time steps (for further explanations, see Menges et al., "How to Make Injection Molds", Carl Hanser).

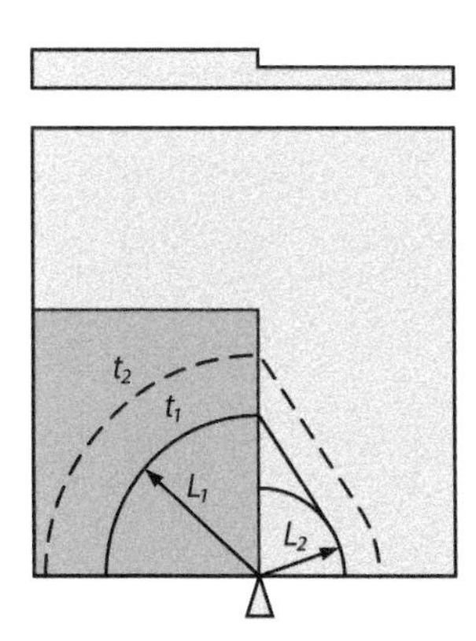

Figure 4.4
Graphical determination of filling progress

The prerequisite is constant viscosity, which is a given at high injection speeds. At low speeds, the melt in the thinner areas will cool more quickly because of the lower flow velocity, the viscosity will increase, and the simple formula will no longer apply.

Very high reliability of the filling calculation

The use of this graphical method for the design process is, of course, outmoded. Simulation programs do this job very reliably and they can also factor in the effect of temperature on viscosity. The filling simulation is shown here in a very simplified way. In fact, prediction of the filling process is quite undemanding and very reliable.

The important results for the designer are:

Prediction of weld lines

- Predicting the *location of weld lines*. Weld lines do not only occur at confluence points behind cores, which the melt must flow around. Even when melt is leading in thicker areas, confluences can occur with adjacent flow fronts. The location of the confluence can be predicted reliably by a program, but not the effect on the optics. For this purpose, software programs interpret the calculation results for the flow front temperature and the angle at which the flow fronts meet.

Air inclusions

- Prediction of *air inclusions* that can lead to the diesel effect (combustion). If melt fronts do not converge at a parting line, the air cannot be displaced. The position of the air inclusion can be predicted precisely, allowing the designer to change the filling pattern by making minor corrections to wall thicknesses. Alternatively, the mold maker can provide measures for venting.

Sink marks

- The location of *sink marks* can also be indicated by a program, although this result is less reliable than the filling pattern itself. Basically, prediction relies not only on knowledge of the local shrinkage potential, but also on the stiffness of the component surface, because shrinkage is more likely to occur at high stiffness. It should be borne in mind that the effect, i.e. the depth, of a sink mark can only be given with a certain probability.

Calculation result for local melt temperature

- The calculation of the melt for *"temperatures over the cross-section"* is interesting. It is possible to display only component areas that have a temperature above the yield point temperature depending on the cooling time. Particularly in the case of components with different wall thicknesses, it can happen that these even warmer areas are cut off from the melt supply via the gate. The result is that holding pressure is no longer possible and there is therefore a risk of sink marks.
- Also interesting is the result for *"local flow front temperature"*. The melt can lose heat during the inflow and so flow fronts merging together will not be able to weld properly. Simulation affords a way to reveal such temperatures, which are not accessible by measurement. Armed with this knowledge, the plastics expert can understand why an area with a weld line might be a mechanical weak point.

4.1.2 Warpage Prediction

Warpage is caused by internal stresses

Predicting the warpage of a component after demolding is very important. Warpage always has to do with stresses in the component that deform the component if the component is not stiff enough. For reliable prediction, the stresses and the component stiffness must be known. The stiffness of a plastic component depends on Young's modulus, which increases steadily during cooling. A prediction of the distortion as a function of the demolding temperatures is not possible with a constant Young's modulus; for this, the simulation program must contain the temperature-dependent Young's modulus function.

Locally different shrinkage

Stresses have different causes. Especially important here is locally different shrinkage, which is only known after the holding pressure phase has been calculated. Warpage analysis therefore requires a calculation of the holding pressure phase in addition to the simple filling analysis.

Stiffness due to fiber orientation

The stiffness of fiber-reinforced plastics depends on the flow direction of the melt. Roughly speaking, the fibers align themselves in the direction of flow, which increases the stiffness of the component in this direction. Warpage prediction therefore requires calculation of the fiber alignment.

Warpage simulation is based on very many model assumptions

The designer should treat the calculation results with caution because their reliability is inevitably lower than is the case for the filling analysis due to the multiple model assumptions made. Every model is either incomplete or addresses only a part of reality. The more precisely a process has to be simulated, the more precisely all the possible influencing variables have to be modeled. Finally, calculation of a very precise model would need a lot of input data and would probably require very long computing times.

Calculation only with accurate material data

For a simulation, the input values should in any case be as accurate as possible. If, for example, the specific material data for the specific application is not available, it is not enough to simply use the material data of what is presumed to be a similar material. A wrong start is bound to produce incorrect results.

Consequence of the warpage analysis

The benefit of the results for the design process are:

- A change in component shape should result in a component with lower warpage upon further simulation. The designer can influence the stiffness of the component by:
  - Making local corrections to wall thicknesses,
  - changing the flow direction by relocating the gating point,
  - changing the fiber orientation (if fibers are present) by local flow braking,
  - providing additional ribbing or stiffening by, e.g., crowning flat surfaces.
- In principle, the designer can *pre-warp the part* by transferring the results of the simulation to the part design with a negative algebraic sign. In principle, the warpage will straighten the component in the process. This approach requires a great deal of experience of interpreting simulation results, and courage because the specifications must be implemented by the mold maker. Thus, the responsibility for success or dimensional accuracy passes directly to the designer.

Warpage simulation also possible without considering the mold temperature

In principle, the mold temperature has an influence on the warpage of an injection molded part. However, the influences of part design are greater, and so there are no major advantages to be gained from a consideration of the actual mold temperature. In addition, it must be considered that the data in the simulation often do not reflect the production conditions.

For the simulation it is therefore reasonable to assume a constant mold temperature and for the mold designer to consider measures for achieving these ideal conditions as far as possible. Under this assumption, a calculated warpage depends solely on the part shape and the flow arising from the chosen gate position. All design measures that result in lower calculated warpage are very likely to result in lower actual warpage in the real component.

4.1.3 Heat-Flux Analysis

Cooling time estimation

In principle, the cooling calculations that form part of the rheological simulation are a secondary result. Here, the temperature of the plastic, especially in the melt area, influences the flowability and thus also the effect of the holding pressure. The calculation of the component temperatures during cooling has no significance for the designer. At most, an attempt can be made to estimate the possible cooling time, but this is not very reliable unless the stiffness of the component as a function of temperature is also known.

Influence of mold design on cooling time

It is important for the designer to know what effect the structure of the component has on the shape of the mold. This concerns the heat flows within the injection mold. The mold should have the same temperature everywhere at the cavity surface. When hot melt hits the cavity wall, the heat from the melt must be dissipated so as to prevent a rise in cavity temperature. Heat flow only occurs when temperature differences are present, and here the focus is on the temperature differences inside the mold, between the temperature control system and the cavity surface.

Parts of the mold, especially tall and slender cores, can heat up a lot during cyclical operation, making the cooling time longer. If a slender core becomes hot over a large area, heat cannot be dissipated there if the temperature differences are too small for the overall hot mold area. The question is therefore how to targeted heat flows may be controlled or how targeted temperature differences generated in the mold without significantly changing the cavity temperature.

Temperature calculations for the mold require cycle calculations

A prediction of the temperature distribution in the injection mold requires that calculations be performed over several successive cycles; this is also referred to as a transient calculation, in which temporal changes in temperature are considered. Due to the limited thermal conductivity of mold steel, a mold heats up during the first approx. 20 cycles and then usually enters a quasi-steady state range over which the temperature curve from melt contact to the next cycle is always the same.

Cycle calculations are reliable even without flow simulation

Because steel has a much higher thermal conductivity than plastic, the temperature changes in the mold are faster and because the injection process is very brief relative to the total cycle time, it is possible to simplify the injection process by specifying spontaneous charging with the appropriate amount of melt of a uniform melt temperature. This form of simulation is then independent of the flow simulation and can also be done with non-rheology software programs, e.g. multi-physics programs.

4.1.4 Calculation of Mechanical Stability (Structural Mechanics)

Structural mechanics concerns the mechanical resistance to load

The calculation of deformation under load requires FEM programs. Young's modulus is a necessary material parameter for this.

- In principle, the value for Young's modulus derived from the standardized tensile test conducted at room temperature does not have to be selected. What is needed is the Young's modulus at the service temperature, which is determined from the modulus of elasticity-temperature curve produced by a DMA test (see Section 5.2.5).

Young's modulus for fiber reinforcements is different in x- and y-direction

- If fibers are used, the Young's modulus values obtained in the different component directions differ substantially, depending on the orientation of the fibers. The values for the local fiber orientations can be estimated with a filling analysis and transferred to the FEM program.

"Mapping" of Young's modulus values

- Two software programs have to be used for the structural mechanics calculation. A rheology program yields the direction-dependent Young's modulus data for all areas of the component. The data are then transferred to a FEM program in a process known as mapping, so-called because the surface is seen as a map onto which the Young's modulus values are transferred. The two software programs basically work with the same component surface, although the meshing of the component is different for each program.

■ 4.2 Base Models for the Rheological Simulation

Simulation result is based on models

A calculation result is only as good as the underlying model. The simulation of the injection molding process is a rheological flow simulation, whereby the term "rheological" refers to the specific characteristics of a plastic. In principle, a simulation needs different models to describe the component shape, the material behavior, and the physical relationships (calculation model). The following information is intended to provide the user with some background to the calculation programs.

4.2.1 Shape Models

CAD models determine the shape of the component

Components are usually designed three-dimensionally. CAD programs record the individual work steps in structure trees, so that individual work steps can be changed again and again. Downstream work steps can, for example, take into account a changed diameter and modify and rebuild the component. This reference is essential, because in the combination of CAD and rheology calculation, the designer will make changes to the design on the basis of, for example, a warpage analysis.

Meshing of CAD data

The problem is that the flow simulation cannot use the CAD data directly. A mesh of area elements or a space grid of volume elements is necessary, in which a filling process is calculated in numerous increments from element to element. The results of the simulation cannot be transferred directly back to a CAD program.

Fine structure specifications lead to very large meshes

Meshing of the CAD data is automatic and initially requires little specialist knowledge. However, the designer should be aware that a very detailed structure has to be represented by more elements than a simple structure. This applies to radii, lettering and possibly also surface texture. These fine structures do not have a major impact on the result of the injection molding process or for the simulation. If possible, the CAD representation of the part for the simulation should therefore be kept simple in this respect.

Transfer formats between CAD and rheology program

It is important that the CAD data be transferred in a format that can be understood by the simulation program. Of the various exchange formats available, the designer should choose STEP, if possible. CAD programs mainly transfer volume data in this format.

Other interface formats transfer only the surface, in which case the susceptibility to errors is greater. In the event of a transfer error, for example, a barely visible gap may be created between two contiguous surfaces, which prevents “overflow of the melt” from element to element during the calculation. In the case of volume data, the meshing is superimposed on the volume, so to speak, with errors in the volume itself being overlooked.

Meshes vary with the type of calculation:

- 1-dimensional for the representation of e. g. cooling channels or cold runners
- 2-dimensional for rapid-filling simulations
- 3-dimensional for warpage simulations

1D meshing for hot and cold runners

The following very brief depiction is not complete and is only intended to give a quick overview. *1-dimensional* means a line between two points. Such lines can be assigned properties, e. g. the cross-sectional shape "round", which signifies a cylinder. Furthermore, it can be assigned the property of "cooling channel" which in turn can be given a definition, e. g. that water flows in it and that heat transfer to the neighboring mold is possible.

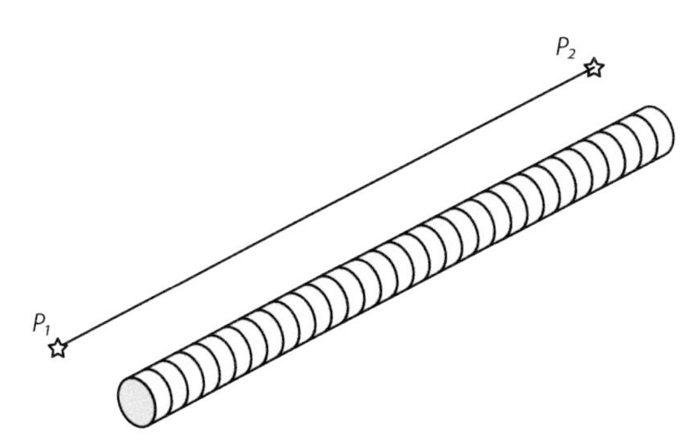

Figure 4.5
1 D simulation model with assigned cross-section "cylinder"

A cooling channel is routed through a mold in three dimensions, but the representation for the calculation can be sufficient in 1 dimension; a change of direction has no bearing on the flow of the cooling medium. For modeling purposes, a cooling channel is created as a line, which is divided into many elements. If the number of elements is too small, large calculation errors may occur.

This model form is also used for cold runners, e. g. when two cavities are to be connected with a common gate. The information yielded by the calculation is not as meaningful as that afforded by the following more elaborate model forms, but the interest here is on the component, which means that a simplification in the runner is often a forgivable error.

Some software allows 1 D models to be used in combination with 2D models.

2D meshing (midplane, surface mesh)

2-dimensonal meshes are often composed of interconnected triangular faces. The individual surfaces or groups of surfaces are assigned properties:

- *Material properties:* E. g. yield point temperature, heat capacity, thermal conductivity, etc.
- *Boundary conditions:* E. g. heat transfer coefficient, material property of the environment (usually steel with heat capacity and conductivity, temperature etc.)
- *Component thickness:* Groups of elements can be assigned the same thickness. In this way, scenarios with different wall thicknesses can be calculated without the need for a detour via CAD data modification

These networks are distinguished as follows:

- *Midplane meshes* represent the components as surfaces without thickness, they are additionally assigned the "thickness" property. A plate is only represented by the center surfaces, which is in the middle of the thickness of the component

(Figure 4.6). Strictly speaking, however, it does not matter whether the surface represents the exact center of the component because, at changes of cross-section, the center surfaces are directly adjacent to each other in the model. One advantage of center surface meshes is that the wall thickness can be changed by simply overwriting the assigned value for the wall thickness with new information.

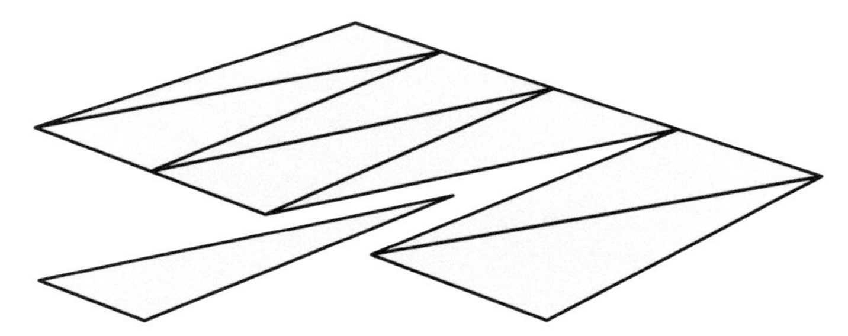

Figure 4.6 2D simulation model "midplane"

- *Surface net:* Here, the thickness of the component initially results from the distance between surfaces that are largely parallel with each other. This can be overwritten with automatically generated information and the model can thus be made thicker or thinner irrespective of how it looks on the screen.

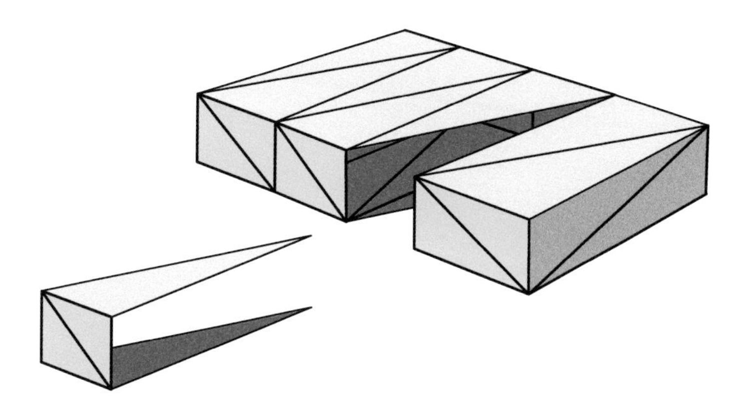

Figure 4.7 2D simulation model "surface net"

Surface meshes are more easily available, require higher computational effort

For center surface meshes, the component would first have to be constructed from planes. Although automated conversions of 3D CAD data into a center-surface model are possible, a certain susceptibility to errors cannot be ruled out. Because the initial data is in 3D form, surface meshes are used in the main.

A disadvantage of the surface meshes compared to the midplane meshes is the longer computation time, even if the computational power of modern computers increasingly reduces this disadvantage. The progress of the filling process of a component is calculated from element to element. In the case of a surface mesh, two filling processes are inevitably calculated simultaneously because a component area consists of a top and a bottom surface. The two calculated flow fronts must be synchronized in a further calculation step, because the melt cannot flow faster on the upper side than on the opposite side in a component area.

As is the case for 1-dimensional meshes, each element first has a result value which indicates whether the element is filled with melt and the temperature of the melt. Further information about the temperature distribution over the cross-section varies with the type of calculation model employed (see Section 4.2.2).

3D mesh is composed of solid elements

3-dimensional meshes consist of volume elements, either tetrahedra (quadrilateral elements made of four triangles joined together) or cuboid elements. With these models, the calculation can also compute processes inside the cross-section of a component, and this is particularly advantageous for thick-walled or bulbous components. Specifically, swelling processes can also be calculated. Experience has shown that the advantage of calculations performed with this type of model is that the results of warpage simulations are more reliable.

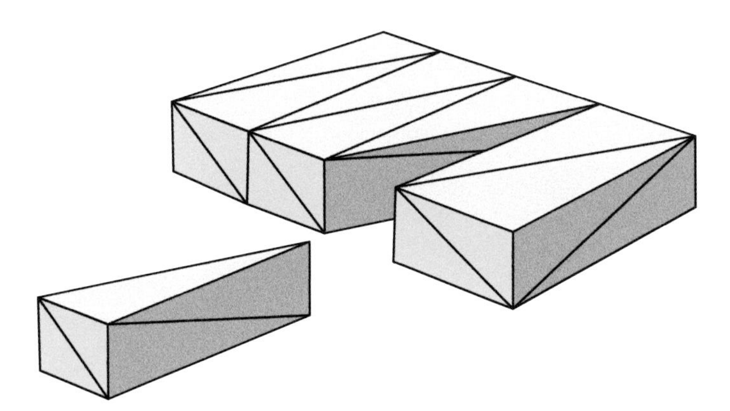

Figure 4.8 3D simulation model composed of tetrahedral elements

In general, for this mesh form, too, each element will have only one result value. If high-precision information is now needed about the cross-section of a component, in principle about 10 elements would be needed to represent the cross-section, e. g. to represent solidified edge layers. Thus, these models become very large in terms of the number of elements to be computed, making the computation time longer. Computers with a very large working memory are then very useful.

4.2.2 Calculation Models

Understanding possible calculation errors

The following considerations are not intended to scare the user, but to provide some understanding of the complexity of the calculation. The aim here is to generate a healthy skepticism about the calculation results.

A rheological calculation is based on three basic physical equations:

- *Conservation of momentum:* This essentially concerns the action of forces
- *Energy equation:* For calculating heat flows to yield concrete temperature data
- *Continuity equation:* Which states that mass cannot disappear

Dilemma of the exact viscosity specification

Conservation of momentum also means equilibrium of forces. Here, the necessary pressure is calculated so that the melt flows at a certain speed. The counterforce is the melt resistance, which depends on the viscosity of the melt. The viscosity depends on the temperature and the shear rate, and so there is a dilemma in the calculation. On one hand, a temperature must first be assumed for a calculation step and thus the viscosity must be known. On the other, only once the calculation has been done is the magnitude of the shear forces known and thus how much heat is supplied via friction. A heat balance equation for the heat supplied in this way and the heat dissipated via the mold provides the temperature.

Filling simulation requires two interlinked calculation steps

For calculations involving 2-dimensional meshes, two coupled calculation steps solve this dilemma. In the first step, the filling progress is calculated at an assumed temperature. The subsequent second step calculates the temperatures for the melt at rest, dividing the thickness assigned to the element into virtual layers (Figure 4.9).

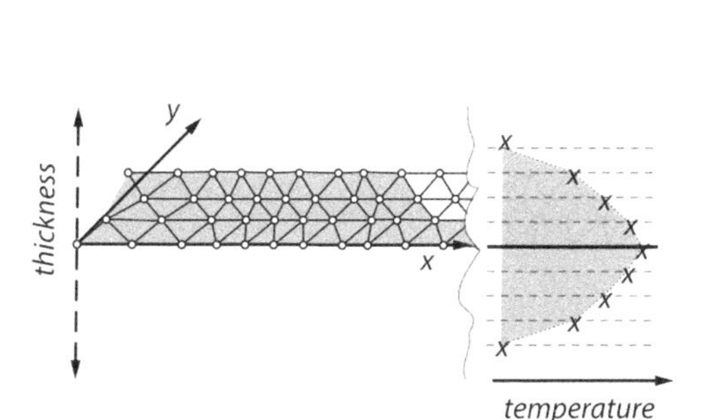

Figure 4.9
Filling and temperature calculation take place in two successive steps

Repeated calculation steps until an error (deviation) occurs is acceptable

For the first calculation step, the flow progress or the pressure and the flow velocity are calculated for each already filled element at an initially assumed temperature. Gradually, the adjacent elements that are still empty are filled, i. e. included in the calculation. When this calculation step has been done for all elements, the temperature of the element is calculated in a subsequent step. If the deviation between the assumed and calculated temperature is too large, the filling process is repeated for this time step until the error is small enough.

Fine meshes show differences in results at high resolution

The user has an influence on meshing. A very coarse mesh results in a large relative error (Figure 4.10). If, for example, a temperature difference of 5 °C over a flow path with a length of 100 mm were mapped with only one element, there would only be one uniform temperature over the entire length, because a calculation result always applies to one element at a time. Finer meshing inevitably produces more calculation results and better agreement with reality, but leads to greater calculation effort with longer calculation times. The error becomes only insignificantly smaller with very fine meshes, so one cannot say that the results are more accurate with very fine meshes. The results initially only show a higher resolution. A concrete statement about the fineness of the mesh cannot be made as a general rule.

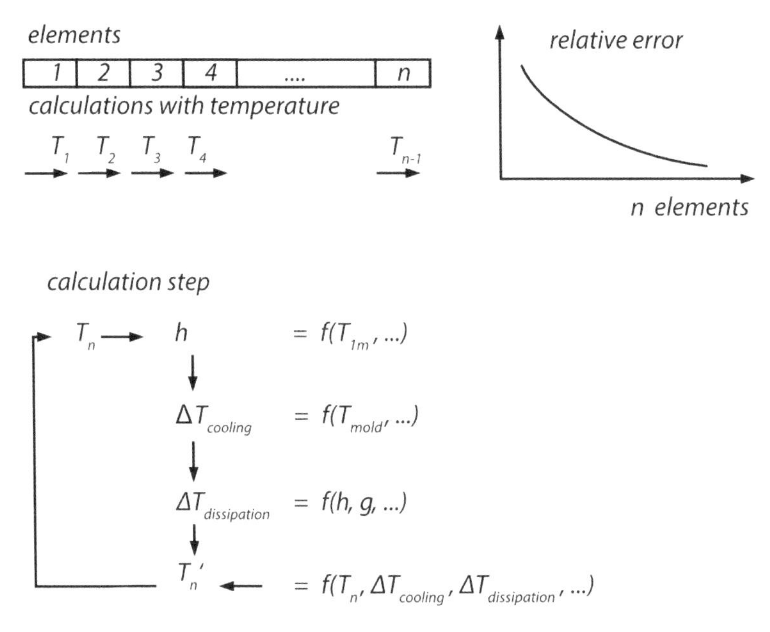

Figure 4.10 Calculation with temperatures assumed in advance

Temperature calculation via virtual layers

The temperature calculation uses a fast FDM calculation method for the 1- and 2-dimensional meshes, in which the assigned component thickness is broken down into virtual layers above and below the center surface. In principle, the wall thickness is only present as a numerical value in these mesh types. For a wall thickness of 2 mm, the simulation program multiplies the respective area and calculates the temperature on the resulting areas. Thus, the program can specify temperatures in the center of the cross-section or just below the surface. The user can influence the number of layers in many programs.

4.2.3 Material Models

Material models describe the properties as a function

Material specifications are necessary for simulation calculations. In the case of plastics, the material data are usually not constant, but dependent at the least on the temperature. The mathematical description of the material behavior as an equation ultimately constitutes a model.

The viscosity is influenced by the temperature and the shear rate. These dependencies can be recorded with measuring instruments. The measurement results are processed and ultimately represented as a function, so that the calculation program receives a concrete numerical value for each temperature and shear rate.

Material models determine the accuracy of simulation results

When choosing the material for a calculation, the user should be aware that bad material models cannot lead to good calculation results.

- If a given material specification or material cannot be found in the program's database, it is highly probable that a similar material will at best exhibit similar material behavior.
- Material data are predominantly only available for the non-colored material. Pigments and dyes influence the flow behavior, so that calculation errors must be expected here as well.
- When the material is being selected, the table data from the program's database should be consulted. If the material specifications are not available as a function but only as a constant numerical value, deviations between simulation result and a real behavior are very probable.

Important material specifications for the simulation

The important material data needed for a filling and holding pressure simulation are:

- Heat capacity and thermal conductivity, usually as constants (1-point values)
- No-flow temperature: Below this temperature, flow is no longer possible, i.e. holding pressure is no longer effective.
- Viscosity as a function of temperature and shear rate
- pvT behavior, dependence of specific volume on pressure and temperature

pvT behavior to indicate the shrinkage volume

The pvT behavior describes the specific volume (reciprocal of the density) as a function of the prevailing pressure and temperature. It is critical to the calculation of the shrinkage volume. For semi-crystalline materials, there is a high risk of error here, which is hard for the user to recognize or assess. The cooling rate affects the crystallization kinetics, which is the speed of the crystallization process. At a high cooling rate, the degree of crystallization is lower. Currently, there are no suitable models that describe the crystallization kinetics. If such models were available, the next challenge would be to have measuring equipment and to generate the necessary material data.

4.3 Examples and Calculation Results

Providers of filling simulation software

Various software programs are available for rheological simulation. Some of them allow only 3D calculations and require powerful computers.

- Cadmould (Simcon, Germany)
- 3D Timon (Toray, Japan)
- Moldex3D (CoreTech System, Taiwan)
- Moldflow (Autodesk, USA)
- Sigmasoft (Sigma Engineering, Germany)
- Rem 3D (Transvalor, France)

Software modules

The software is predominantly modular, i.e. the buyer can purchase additional modules depending on his/her requirements.

Table 4.1 Software Modules

Flow	Filling phase, calculation of filling behavior, detection of weld lines and air inclusions
Pack	Holding pressure, calculation of local shrinkage, detection of sink marks
Fiber	Fiber alignment, calculation of local alignment of fibers when using fiber reinforced materials.
Shrink/Warp	Shrinkage/warpage, calculation of the warpage
Cool	Cooling phase, calculation of local temperatures at the cavity surface based on further model information on the mold structure and especially the position of temperature-control channels
Stress	Stresses, stiffness calculation, also buckling and bumping

4.3.1 Filling Behavior

Filling simulation for selecting the appropriate gating point

The filling analysis provides the filling pattern, which can be used to detect weld lines and air inclusions for selected gate positions. The designer can then decide to relocate the gate or to influence the filling behavior via wall thickness changes (flow brakes, flow aids). If the melt flows together and encloses areas that are still unfilled, the air cannot escape through the parting line here and local burning will occur (diesel effect). Possible solutions are:

- Mold venting via an ejector pin at this point
- Wall thickness reduction laterally and parallel to the emerging weld line, so that the flow front advances more slowly here

- Wall thickness increase on the area before air entrapment, so that the melt reaches and fills this area sooner, before the weld line is formed

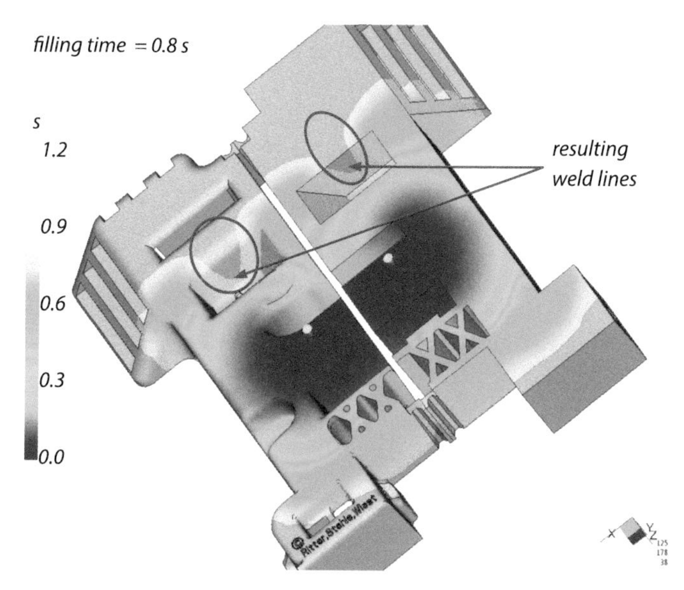

Figure 4.11 Weld line and possible air entrapment at the end of the filling process

Velocity vectors reveal cause of differential shear heating

Velocity vectors reveal interesting information. Each element has a value for the flow velocity at each point in time. A look at these velocities shows where the melt is flowing more slowly. This provides important information because a slow-flowing melt, in particular, experiences less shear heating and thus cools faster and becomes more viscous. This influences the filling process and the weld line behavior. Figure 4.12 shows that, at the filling time of 1.1 s, the melt in the lower area has already almost come to a standstill and also shows the areas in which the melt can still flow from the two sprue points at a higher velocity.

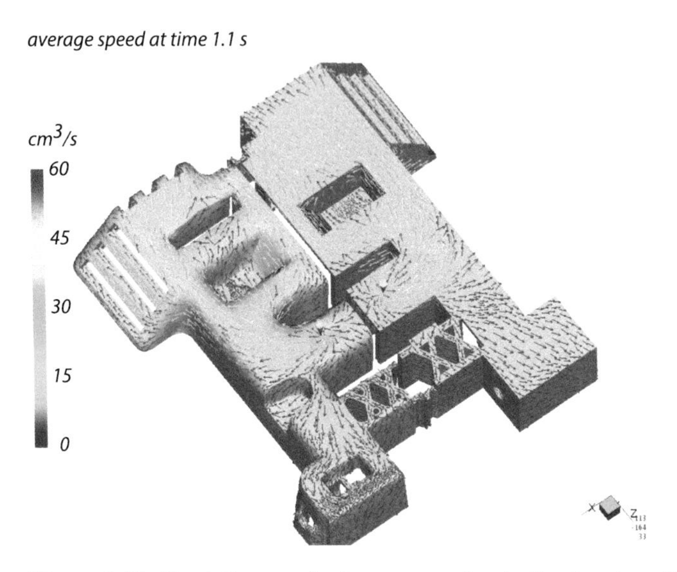

Figure 4.12 Simulation result of average melt velocity at a given filling time

Flow front temperature for judging possible weld line weaknesses

The calculation result for "flow front temperature" shows the temperature of the melt at the time when it reaches the surface at the flow front. Figure 4.13 shows that the melt at the end of the flow path has become approx. 4 °C colder. There is also a noticeable temperature difference on the left side of the component, where a weld line is formed. If low temperatures only prevail in weld lines, the strength of the weld line will be poor and, in all probability, it will also become more noticeable.

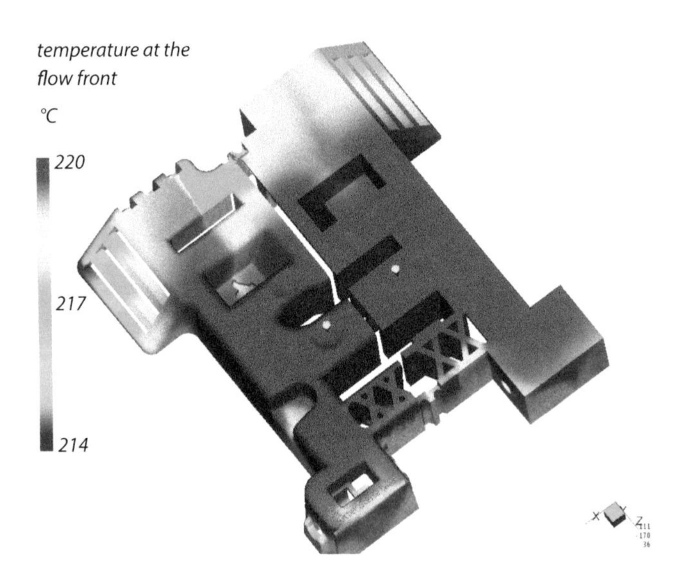

Figure 4.13 Calculation result for "flow front temperature"

4.3.2 Holding Pressure Phase

Calculation of holding pressure phase for shrinkage volume specification

Simulation of the holding pressure phase provides results on the temporal change of the melt temperature after injection and thus the local shrinkage volume. The holding pressure varies in the component, and at some point, at the end of the flow path, the holding pressure is no longer effective due to the increasing viscosity of the melt. This means that no further melt can be conveyed there to compensate for the further shrinkage.

Display of still active holding pressure areas only with 3D calculation

A 3D simulation offers the possibility of visualizing elements whose temperature at a given time lies within a selected temperature range. Figure 4.14 shows all elements that are colder than the yield point temperature of the material (here: 130 °C) after 7.3 s. It can be clearly seen that, in the upper part of the image, larger volume areas are still flowable, but the areas behind the sprue have already frozen off. Therefore, no holding pressure can compensate for shrinkage here. Sink marks can be expected.

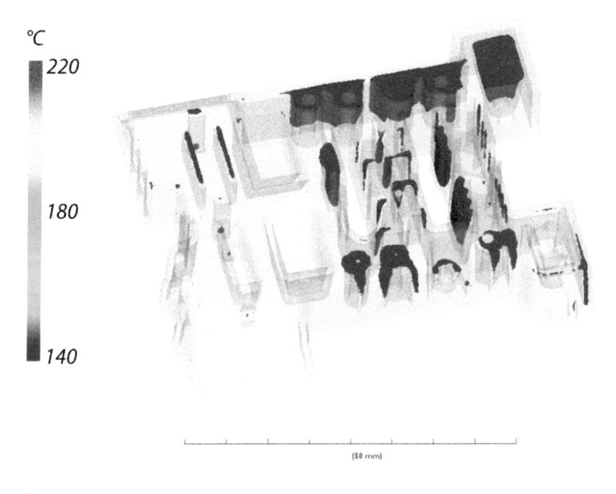

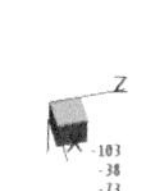

Figure 4.14 Simulation result for "areas not yet frozen"

Information about the extent of sink marks is interpreted by the software

Figure 4.15 shows a forecast of the possible sink marks. Generally, it is important to realize that such predictions of sink marks and weld lines are not necessarily accurate. Whether a sink mark occurs depends not only on the possible shrinkage potential, but also on the stiffness of the component surface. In the case of a stiff surface, the shrinkage potential can also cause a shrinkage cavity below the surface instead of a sink mark. The calculation result depends on assumptions and mathematical calculations.

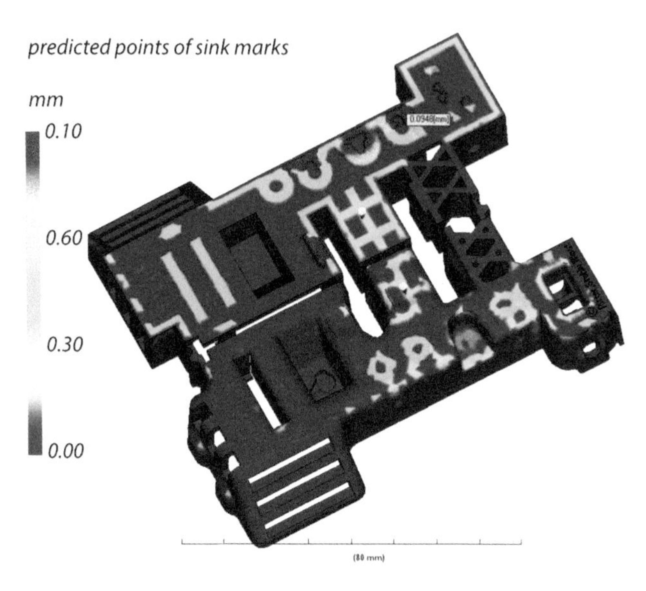

Figure 4.15 Simulation result for "predicted sink marks"

Taped seams, depending on the angle of confluence and temperature

The same applies to the calculation result for "weld lines." The visibility of a weld line depends on the local temperature of the flow front and the angle at which two melts meet. The calculation result in Figure 4.16 shows the calculated possible weld lines and the angle of confluence in each case. In this case, an angle of 0° means that two flow fronts meet head-on. For angles shallower than 135° the program does not output any weld lines.

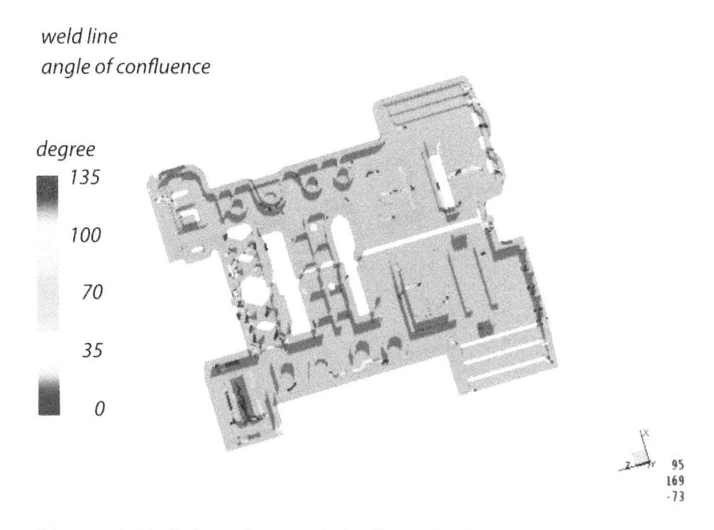

Figure 4.16 Calculation result for "weld line"

The designer should rely less on the display of weld lines and sink marks and instead carefully monitor the filling curve over the injection time while additionally paying attention to the information about the calculated temperatures.

Automatic data should be questioned

4.3.3 Warpage

The warpage calculation evaluates the calculated stresses after the end of the holding pressure phase. The designer should be aware that any model can only be a partial representation of reality. If several models are linked together and a calculation is started with the result values obtained from a first calculation, the results should be used with caution. The result is a graphical representation of the warpage in three dimensions or the total warpage. It is possible to scale the representation, by which is meant that the warpage is visually amplified by a defined factor. This affords a quick way of determining the direction in which dimensional deviations from the desired component shape are to be expected.

Warpage result is the result of filling simulation and subsequent holding pressure simulation

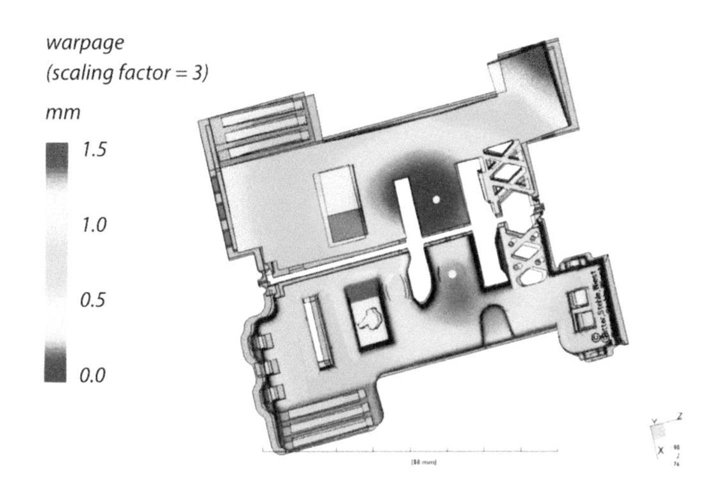

Figure 4.17 Simulation result for "warpage"

If selected component points are defined as anchors, the distortion is displayed relative to these points. This is advantageous, because a component of an assembly is also anchored at the interfaces to adjacent components and dimensional deviations at distant component points can be detected more clearly.

5 Material Selection

Questions governing the right choice of plastic material

The right choice of plastic is made according to various aspects:

- Technical reasons: Transparency, temperature application, stiffness, etc.
- Experience: "We've always done it that way!"
- Legal regulations:
  - Time-consuming regulations for approval, e.g. aviation or medical technology
  - Approval requirements, e.g. for contact with drinking water and foodstuffs

In the following, a system is presented which facilitates the selection of materials. The characteristic values mentioned differ substantially from those for metals. They are therefore explained in detail after the systematic filter approach.

Not every development requires extensive material selection. In the case of drinking water and food applications, as well as in the medical and pharmaceutical sectors, approvals for plastics play a cost-intensive role, so that in these industries the material is only changed if significant benefits can be activated. The same applies to materials with which experience has been very good, such as polyamides in automotive engine compartments.

5.1 Usual Procedure for Selecting Materials

The conventional approach to material selection is a filter principle

The choice of plastic is made according to the filter principle (Figure 5.1). The order of the sieve inserts is arbitrary. The order shown here is based on the respective number of materials available. Only a very few plastics can fulfill very special requirements, e.g. transparency coupled with simultaneous high-temperature application. A considerable number are suitable for application temperatures in the range up to 60 °C.

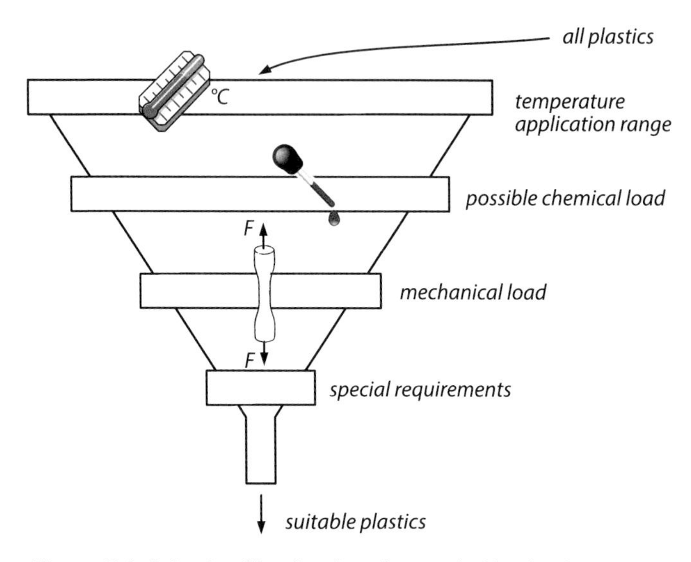

Figure 5.1 Selection filter for choosing a suitable plastic

At each stage of the filtration process, the essential requirements for the application have to be specified. For an application, only those plastics are considered which meet the specified requirements.

5.1.1 Selection Criterion: Temperature

The heat deflection temperature and the continuous service temperature are important filter criteria

When temperature is the selection filter, the upper limit always applies. With increase in temperature, mechanical properties deteriorate, and thermal damage can occur. There is no lower temperature limit. At low temperatures, the material becomes brittle, but this property is subsequently queried in the mechanical properties filter.

For the application temperature, the question of exposure time needs to be answered.

a) **Short-term:** For non-long-lasting high temperatures, the heat deflection temperature (HDT) should be used. This is the temperature at which a standardized specimen does not exceed a permissible deformation under load. As the temperature increases, Young's modulus and thus the stiffness of a component decreases. If there are no significant load requirements for the desired requirement, this temperature limit can also be ignored.

b) **Long-term:** For long-lasting high temperatures, use the continuous service temperature. This temperature indicates that a property has decreased by 50% after an exposure time of 20,000 hours.

Plastics are organic, i.e. they consist of carbon compounds. At elevated temperatures and in oxygen-containing atmospheres, the molecular structure can be damaged; this process is called thermal degradation. Degradation of the material depends on time and temperature. During production, the plastic is molten and has a very high temperature, so the material can be damaged quickly here. In application, temperatures tend to be low, and degradation therefore tends to be slow. Clear signs of degradation are either color changes (yellowing, becoming milky) or brittleness under dynamic load. In this book, the focus is on the field of injection molds. The designer should be aware that his/her design specifications must ultimately be implemented by means of an injection mold.

An unfavorable choice of plastic can lead to thermal degradation and, in the worst case, to component failure

5.1.2 Selection Criterion: Chemical Load

Plastics have a molecular structure. There is plenty of space between the individual molecules for other, smaller molecules, e.g. "chemically similar" liquids. One well-known example is the penetration of dyes into PP containers, for example. The penetrating molecules may cause the plastic to swell, which means an increase in the distances between the molecular chains. This reduces the mechanical properties and can lead to cracks under load.

The chemical load applied by the user must be considered in advance when selecting the material

In many cases, the stress is already present in the material in the form of internal stresses. These stresses cannot be completely avoided in the manufacturing process. Contact of a component with a fluid may therefore lead to sudden component failure.

When selecting materials, it is necessary to check exactly which substances can come into contact with the component. This also applies to cleaning agents that the user may want to use later. It is obvious if the component can be used with acids (e.g. car batteries) or with oil. Databases can be used to quickly filter which plastics are to be used in combination with special fluids.

5.1.3 Selection Criterion: Mechanical Load

The criterion governing mechanical loading capacity is extensive and includes several questions:

Criteria for the limit of mechanical load include:
- Loading temperature
- Young's modulus
- Load type and duration

1. At what temperature does the loading take place?
   - The **loading temperature** must be specified. Young's modulus of plastics depends very much on the temperature and, for some plastics, additionally on the ambient humidity. In the case of polyamide (PA), Young's modulus can be 3,000 N/mm^2 in winter and in dry air, and 1,000 N/mm^2 in a muggy mid-summer.

Operating temperature

- In many cases, Young's modulus obtained from the short-term tensile test is used for the load calculation. This standardized test takes place in a standard climate at room temperature. For applications at 60 °C, the measured value will be too high, depending on the plastic.
- Material selection therefore consists in choosing a material that has the required Young's modulus at a given temperature. For this purpose, Young's modulus-temperature curves are available, which are determined by means of dynamic mechanical analysis (DMA). If these measured values are not available, the value of the short-term tensile test can be reduced by estimation factors and used as a substitute.

Static or dynamic load

2. Is the load sudden and dynamic, i.e. an impact load?
   - The higher the temperature, the more likely a plastic is to be tough to soft. Therefore, sudden impact loads are more critical for use at low temperatures. Impact strength is usually determined at 23 °C. Depending on the application, values for lower temperatures are necessary and must be determined accordingly.
   - For an application, consideration should be given to the conditions under which failure may occur. Latching hooks are predominantly loaded only once during assembly and then, if necessary, abruptly. Applications in automotive interiors (e.g. dashboard) can be loaded at different temperatures in the event of a crash.

Load duration

3. Is the load more short-lived or long-lasting and constant?
   - Most plastic components tend to be lightly loaded. This applies to housings and cladding, for example. If a heavier load is applied, the component will reach its mechanical limit and possibly fail. A clear distinction must be made between short- and long-term loading.
   - Short-term loads are all loads that last less than one hour and can always be repeated, e.g. snap-fit connections during joining.
   - Long-term loads tend to last for days and longer. Where these are to be expected, the highest possible probable loading temperature has to be specified in any case. This is because the plastic can creep under such loads.

Creep is a phenomenon that must be monitored in semi-crystalline plastics

   - Under a permanent load, a plastic can creep, which can take the form either of steady strain or of compression. The higher the temperature of the load, the greater this deformation is.
   - Creep occurs predominantly at temperatures above the glass transition temperature T_g. Therefore, the amorphous plastics, which are always used below T_g, are advantageous. Another limit value for amorphous plastics is the permissible critical strain of 0.5%. This value is not a characteristic value in the true sense, but a proven value in practice. Amorphous plastics that perma-

nently tolerate a strain of more than 0.5% form crazes. These are primarily visual defects that appear as fine hairline cracks. The mechanical properties are not significantly reduced when crazes occur.

4. Is the load one that recurs cyclically or fluctuates permanently?
   - Unlike metals, there is no **fatigue strength** under cyclical loads. In the case of metals, local hardening and subsequent fatigue fracture occur at load changes below the yield strength ($R_{po,2}$, R_e). Plastics have a molecular structure and therefore no slip planes. A comparable hardening mechanism is therefore not known.
   - Under cyclical loads, however, self-heating can occur, especially at temperatures above the glass transition temperature T_g and at higher frequencies, and reduces Young's modulus. This affects the semi-crystalline plastics which are also used above T_g.
   - In the case of fiber-reinforced plastics, repeated load changes can cause damage to the bond between the fibers and the plastic. In the long run, this leads to component damage. However, these processes are still largely unexplored. Measurements and characteristic values are not available here.

There is no fatigue strength for plastics under cyclical loads

5.1.4 Selection Criterion: Special Requirement

There are many special properties, and a complete list cannot be provided here. Some are explained below by way of example.

Some plastics have special properties

- *Colorability:* In principle, all thermoplastics can be colored. The polymer raw material is almost always milky opaque in the case of semi-crystalline plastics and largely transparent in the case of amorphous plastics. The coloring effect of the dyes or pigments depends on the basic coloration of the raw material. The more transparent the plastic is, the more the color of the interior is also visible.
- *Dimensional stability:* Dimensional stability is an important criterion for the manufacturing process. Since plastics exhibit pronounced thermal expansion, they shrink during cooling. This can lead to undesirable changes in shape. Fillers can be used to reduce this shrinkage behavior. Among those known here are glass fibers which, depending on their position and arrangement in the component, produce different properties in the longitudinal and transverse directions. Glass beads of a few 1/100 mm in size are preferable here, because they maintain largely isotropic properties (the same in all spatial directions).
- *Surface finish:* Scratch resistance is a very important surface property. Some plastics (PS, PMMA) have a sufficiently hard surface. If necessary, this requirement can also be generated subsequently by means of a UV-curing coating.

- *UV resistance/weather resistance:* UV radiation can damage a plastic over the long term. The result of this is molecular degradation which, depending on the color of the plastic, is accompanied by yellowing or graying. This damage mechanism can be delayed by incorporating additives.
- *Surface coatability and adhesiveness:* After component manufacture, painting or coating is one of the finishing processes. Paintability is very difficult in the case of some plastics because the paint does not adhere to the surface. The adhesion of paint or adhesives also depends on the surface tension, which varies from one plastic to another. It is possible to improve paint or adhesive adhesion via downstream processes. In principle, however, downstream processes increase the manufacturing costs, so that it is advantageous to choose a plastic that is immediately suitable.
- *Processing properties:* A very important property is the flowability of the melt. Here, again, this property can be influenced over a wide range by incorporating additives. In the case of related plastics, grades of different molecular weights are also available. The higher the molecular weight, the poorer the flowability and the higher is the melt strength. Melt strength is important for extrusion and blow molding processes, but has no bearing on injection molding applications.

5.1.5 Databases

Databases offer the designer a wealth of information

A survey of design engineers shows that about 30% of the information on the plastics to be selected is already available within the company in the form of empirical knowledge, is obtained from raw material manufacturers, and comes from databases. Direct inquiries made of raw material manufacturers are particularly suitable, because special mixtures of raw materials and suitable additives may have been developed for special applications with very specific requirements.

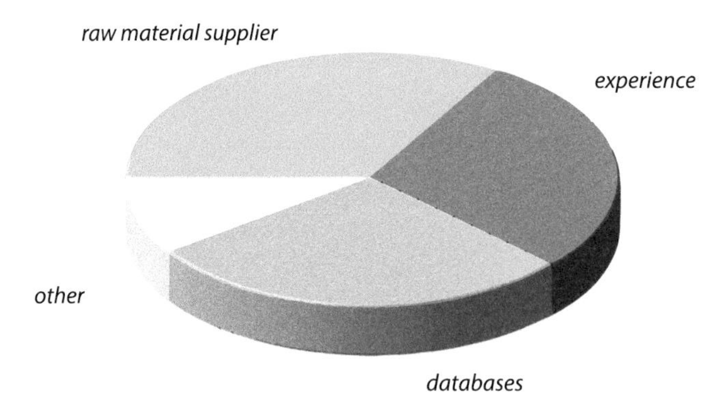

Figure 5.2 Sources of information on plastic properties and material selection

There are several commercial and freely available, non-commercial databases to be found on the internet:

- Commercial
  - M-Base
  - CES-Selector (Granta Design)
  - POLYMAT
- Freely available
  - Campus
  - IDES Prospector
  - The Campus database (Computer Aided Material Preselection by Uniform Standards) is the most widely used materials database in the world. It is available online through some raw material manufacturers. The data structures are uniform for a standardized specimen geometry, tooling, processing, and testing conditions. The material properties correspond exactly to the international standards.
  - The UL Prospector database includes data sheets for around 87,000 plastics and 1200 additives. It contains data on almost all-important polymers from approx. 900 raw material manufacturers worldwide.

The freely available databases Campus and UL Prospector are among the most widely used

Single point data and characteristic value functions

The data available here are single point (ISO 10350) and multipoint (ISO 11403) types. Single point means that only a single characteristic value is available, while multipoint means a characteristic value function. Young's modulus, for example, is a single characteristic value.

However, a conscientious designer needs Young's modulus at selected temperatures and will look for multipoint data, if possible.

For the designer, many characteristic values are of only limited importance

A review of databases shows they contain many characteristic values which are of limited importance to the designer. In the following section, the application for these values is first explained and then helpful background information is given on the values that are important for the design.

In the Campus database, the information on the material data is sorted into groups:

Single Point Data

Single point data are used for a quick overview and selection

MVR or MFI indicates the flowability of the melt

- Rheological values
  - The MVR (melt volume rate) describes the flow behavior of the plastic melt. It measures how much melt volume flows through a defined nozzle cross-section at a specified temperature and pressure. Similarly, the weight of the outflowing melt can also be measured, in which case it is referred to as the MFI (melt flow index). Large values indicate a readily-flowing material. For design purposes, this information is of no significance, even though it is used to limit the maxi-

mum flow path length. However, a direct calculation is not possible with this value; where there is doubt, the filling behavior of a mold should be checked by means of simulation.

Shrinkage

- Two percentage values are given for the shrinkage behavior, namely longitudinal and transverse. Fiber-reinforced materials in particular have different values, depending on the orientation of the fibers to the flow direction. Shrinkage values are not material constants; they apply only to the specified test conditions. Shrinkage can be influenced by changing the process settings (especially the holding pressure level).
- It is not really possible for the designer to apply these values in a meaningful way. The mold maker will dimension the mold cavity on the basis of this information and, if necessary, use a simulation calculation for this purpose.

Young's modulus and tensile creep modulus

- Mechanical parameters
  - The tensile test yields Young's modulus (also called modulus of elasticity) as well as the load limits under controlled conditions.
  - The tensile creep modulus is measured in long-term tests but is only available for a few materials.

Fracture behavior

  - Impact strength is specified using the Charpy notched impact test with notched and unnotched test specimens for values at 23 °C and −30 °C. This value provides information on the fracture behavior after an impact load, i.e. whether the material fractures in a more brittle manner or still deforms tenaciously during fracture. In the case of brittle fracture behavior, sharp fracture edges are very probable.

Glass transition temperature and melting temperature

- Thermal specifications
  - The glass transition temperature is specified for some plastics. It marks the transition between brittle and tough behavior. For amorphous materials, it is the maximum upper operating temperature.
  - The melting temperature is of no importance to the designer. The melting temperature exists only for semi-crystalline plastics, and is the temperature at which the crystallites melt.

Heat deflection temperature

  - Information on the upper operating temperature (for measurement method, see Figure 5.4) if high loads act on the component.

Coefficient of expansion (longitudinal, transverse)

  - Thermal expansion is given for temperature changes from 23 to 55 °C or −40 to 100 °C. In the case of fiber-reinforced materials, data are available for longitudinal and transverse orientations.

Flammability

- Fire behavior data
  - Except for PTFE and silicone, most plastics burn well; they consist mainly of carbon. The burning behavior can be influenced by additives, such that their

ignition is delayed, they burn more slowly overall, or they are self-extinguishing.

- Electrical specifications
  - Information on dielectric strength during overvoltage and on dielectricity.
- Other information
  - The water absorption capacity shows how much moisture a plastic can absorb. This information is very important because moisture absorption decisively changes the properties of plastics. In many cases, the above data are given both in a conditioned and in a dry state. Conditioned means that the plastic has taken on the moisture of the normal environment. The mechanical values are then lower than those of a dried test specimen.

Moisture absorption

Multipoint Data

Multipoint data combine several characteristic values in characteristic value functions

- *Rheological values:* Viscosity refers to the flowability of a plastic melt. It is influenced by the temperature and the shear rate. This behavior is insignificant for the designer; the usefulness of this data pertains exclusively to simulation.
- *Mechanical characteristic values:*
  - Young's modulus is strongly influenced by the operating temperature. With a special measuring technique (DMA, see Section nnn), Young's modulus can be plotted as a function of temperature. Depending on the structure, this measurement technique is also used to measure the shear modulus G. In principle, both moduli are in a direct relationship, and can be estimated as follows: $E = 3\ G$. For the designer, this temperature-dependent value is very important.
  - The creep modulus is plotted as a function of time. If possible, values should even be available for different temperatures. The creep modulus can be used to estimate how much a component will permanently deform under a static load after certain times.
  - The pvT data show the specific volume v as a function of pressure and temperature. The specific volume is given in cm^3/g and is the reciprocal of the density (g/cm^3). Changes in this value thus indicate the compressibility essentially of plastic melts. These values are only important for the simulation.

Viscosity at different temperatures and shear rates

Young's modulus and creep modulus as temperature function

pvT data

A detailed analysis of the large amount of information available shows that, of almost 40,000 database entries examined, around 20,000 contain information on Young's modulus. For the very important information on time-dependent behavior, the hit rate is only about 11%. The reason lies in the simplicity of the short-term tensile test compared to the data determined in the long-term test.

Unfortunately, elaborate multipoint data tend to be under-represented

Tabelle 5.1 Existing Information from Databases [according to: Kunz, J., "Bauteilauslegung mit Augenmaß", Kunststoffe 12/2013, p. 56ff]

Properties	Number of Materials	Share in Percent
Total materials	39,551	100.0
Tensile modulus	18,188	46.0
Yield stress	15,948	40.3
Tensile stress at break	12,355	31.2
Tensile creep modulus 1 h	567	1.4
Tensile creep modulus 1000 h	651	1.6
Isochronous stress-strain diagram	502	1.3
Diagram creep modulus time	446	1.1
Creep curves	454	1.1

5.2 Important Characteristic Values

5.2.1 Characteristic Temperatures

The various plastics have characteristic temperatures. Independent of fillers, these temperatures can be used to distinguish and identify the plastics from one another.

The range of application for amorphous plastics is limited to temperatures below the glass transition temperature. Semi-crystalline plastics can be used up to the melting temperature range; below the glass transition temperature, they tend to be brittle. While the glass transition and melting temperatures only give a rough indication of the behavior, they are two definite temperature limits of relevance to design.

5.2.1.1 Glass Transition Temperature

The glass transition temperature is the temperature at which the behavior of the plastic changes from elastic to viscous

All plastics have a characteristic glass transition temperature T_g or a range of a few °C over which the behavior changes from elastic to viscous or viscoelastic. By viscous is meant time-dependent plastic deformation. Plastics are viscous because of the long, tangled molecular chains. In many cases, the melt behavior is described as plastic, although here, too, the behavior is time-dependent.

Elastic property does not directly mean rubber-elastic, i.e. highly stretchable. In the elastic state, plastics predominantly range from brittle to hard.

5.2.1.2 Melting Temperature

Crystals melt above the melting temperature T_m, which is why only semi-crystalline plastics have this characteristic temperature. Because large crystals melt at higher temperatures and a plastic has a range of crystallites of different sizes, the melting temperature is always a range spanning a few °C.

Melting temperature concerns only the melting of crystalline domains

Semi-crystalline thermoplastics are viscoelastic above the glass transition temperature. In principle, two phases are present in the temperature range between the glass transition temperature and the melting temperature of the crystallites, namely the crystalline phase and the softened phase. As a result, these plastics are very tough and ductile in this temperature range. In this range, the time-dependent behavior is special, i.e. under load, permanent deformation takes place with increase in time.

Amorphous thermoplastics have no melting temperature because they contain no meltable crystallites. These plastics increasingly soften above the glass transition temperature and are ultimately flowable.

It is important here to distinguish between melting temperature and melt temperature. At high temperatures, a thermoplastic becomes a liquid which is generally referred to as the melt, whereas amorphous plastics reach this state via continuous softening above the glass transition temperature and not by melting.

Difference between melt temperature and melting temperature

5.2.1.3 Degradation Temperature

The upper temperature is the degradation temperature T_d. Even in the absence of oxygen, the very long molecules are no longer stable here and break down into smaller molecules. The plastic decomposes and loses its properties. The decomposition temperature is not characteristic of a plastic, because thermal decomposition is strongly dependent on temperature, time, and atmospheric oxygen. More suitable for estimating the upper stress temperature is the continuous service temperature (see Section 5.2.3).

Degradation temperature – undefined upper temperature limit

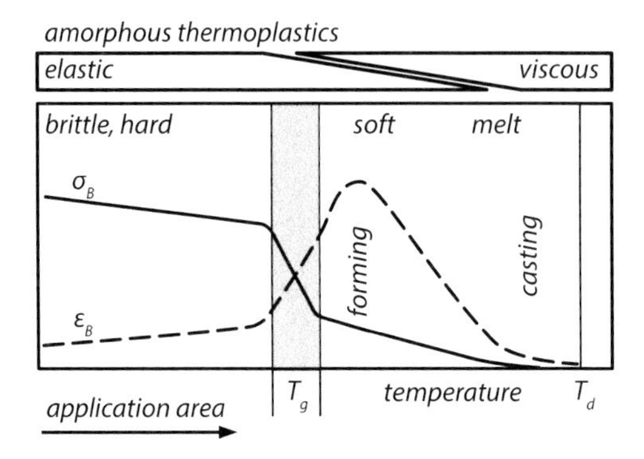

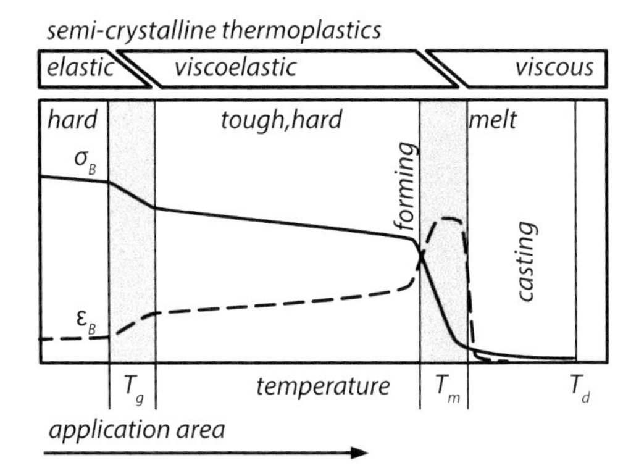

Figure 5.3
Behavior of thermoplastics at different temperatures

5.2.2 Heat Deflection Temperature

Heat deflection temperature is the limit of mechanical strength

The heat deflection temperature indicates the temperature limit of the mechanical load-bearing capacity. In this case, the load-bearing capacity is sufficient stiffness. Young's modulus becomes increasingly lower at higher temperatures and with it, the component stiffness.

There are two different measuring methods:

HDT – Heat Deflection Temperature

Measuring method for heat deflection temperature

The measurement is performed in a three-point bending test in a liquid bath with a rectangular test specimen lying flat on two supports and loaded centrally. The test assembly is heated up in a controlled manner via the liquid (Figure 5.4). It is assumed that the specimen always has the same temperature as the bath at the low heating rate. The load is either 1.8 MPa (HDT-A) or 0.45 MPa (HDT-B). The load σ results from:

$$\sigma = \frac{F3L}{2bh^2} \tag{5.1}$$

where F is the load, L the distance between the supports, b the width of the specimen, and h its height.

The resulting value is the temperature at which the specimen is stretched by 0.2% due to deflection Δs in the edge region. This deflection is calculated from:

$$\Delta s = \frac{L^2 \Delta\varepsilon}{600h} \tag{5.2}$$

or for an edge fiber twist of 0.2% as specified in the standard:

$$\Delta s = \frac{L^2}{3000h} \tag{5.3}$$

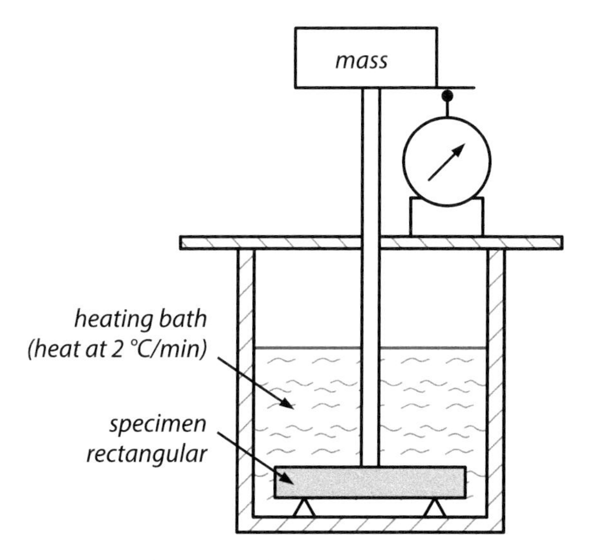

Figure 5.4 Heat deflection temperature measurement

Heat deflection temperature and continuous service temperature

The heat deflection temperature indicates the temperature up to which the mechanical properties are still adequate. Figure 5.5 shows the heat deflection temperature values for material groups in comparison with the continuous service temperature (see Section 5.2.3). Some plastics may still be mechanically suitable at a higher temperature. However, this only applies for a limited period. Other plastics can withstand higher temperatures in the long term, but should not be subjected to mechanical stresses.

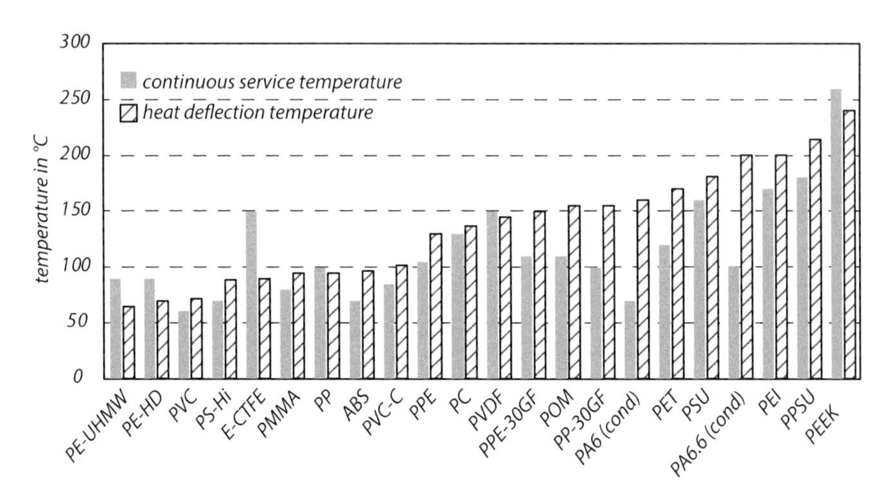

Figure 5.5 Continuous service and heat deflection temperature of common plastics

Vicat Softening Temperature

Softening temperature

The **Vicat softening temperature** (VST) is measured in accordance with DIN EN ISO 306 using a needle (with a circular surface of 1 mm²). This is loaded with a test force of 10-N (test force A) or 50 N (test force B). The specimen with a permissible thickness of 3 to 6.4 mm is heated at a defined rate of 50 or 120 K/h, respectively. The VST is reached when the indenter reaches a penetration depth of 1 mm. According to the standard, the test is only to be used for thermoplastics and provides information about the practical continuous use limit, which is approximately 15 K below the Vicat temperature.

5.2.3 Continuous Service Temperature

The continuous service temperature is the upper temperature limit that plastic components can withstand over a long period of time without noticeable thermal degradation. Degradation is a chemical reaction and becomes faster with increase in temperature. The rate increase is exponential, so at low temperatures it is very slow. At higher temperatures, degradation accelerates.

$$Speed\ of\ chemical\ reaction = Ae^{\frac{B}{T}} \tag{5.4}$$

Here, the values A and B are constants. The natural logarithm of the equation for the reaction speed yields a linear plot of degradation rate against temperature ($1/T$).

$$\ln(reaction\ speed) = A + B\left(\frac{1}{T}\right) \tag{5.5}$$

If it is assumed that degradation is a chemical reaction and if this degradation can be represented as a straight line, extrapolation to higher temperature is possible. For the determination of the continuous service temperature, test samples are stored at different temperatures for a long time. The properties E are measured at fixed time intervals. When the property is only 50% of the original value, the storage time for this value E is noted. These values E line up linearly for different temperatures if $1/T$ is chosen as the axis (Figure 5.6). In this way, for a time of 20,000 h, it is possible to extrapolate to the temperature that leads to a 50% reduction in properties.

Measurement of the continuous operating temperature

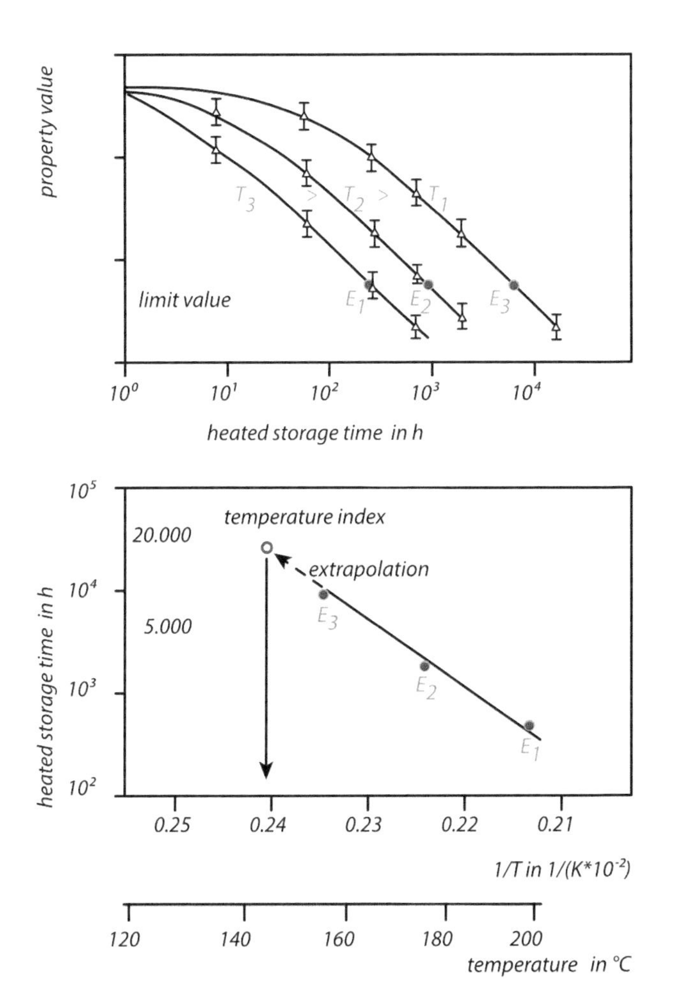

Figure 5.6 Determination of the continuous operating temperature

5.2.4 Young's Modulus and Creep Modulus

Young's modulus or secant modulus

For plastics, Young's modulus is calculated from the initial slope of the stress-strain curve. In many cases, this curve is not linear even at low stress. For this reason, the initial range is defined here as linear over two support points at 0.05% and 0.25% strain, and Young's modulus is thus determined from the slope between these two points. Therefore, this value is often also referred to as the secant modulus.

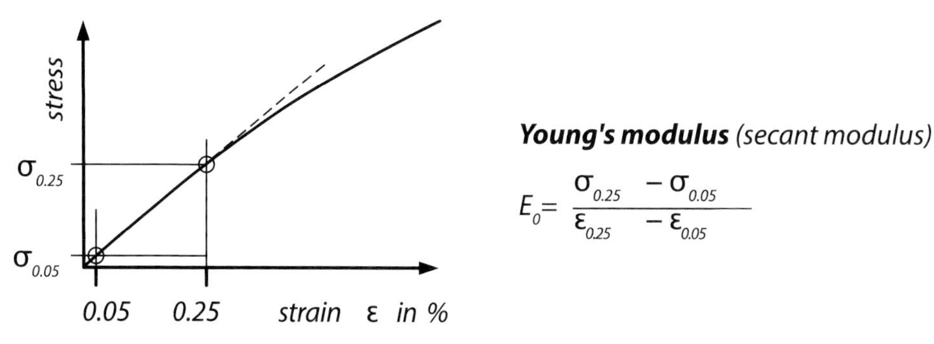

Figure 5.7 Characteristic values of the tensile test

Young's modulus decreases at higher temperatures and higher load speeds

The mechanical behavior of plastics is very different from that of metals. At temperatures below the glass transition temperature, they are largely elastic. At increasingly high temperatures, they become viscous. This means that the results curve for the tensile test becomes increasingly flat, i. e. Young's modulus becomes smaller. At low loading speeds, Young's modulus is also smaller.

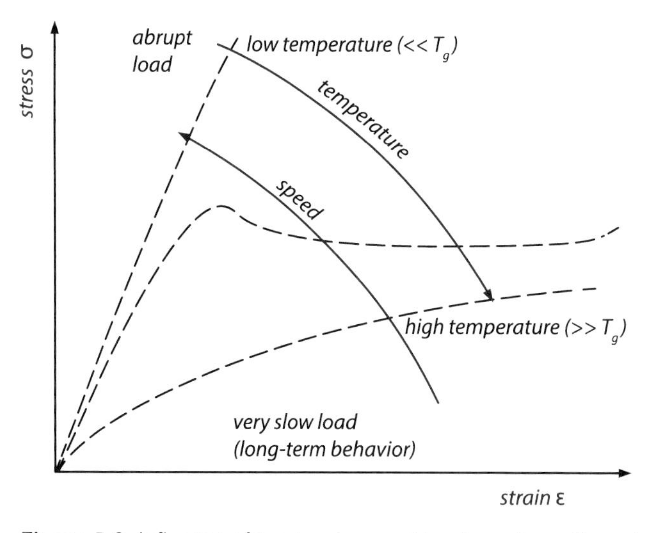

Figure 5.8 Influence of temperature and loading rate on Young's modulus

Some plastics can absorb moisture from the environment. This also leads to a noticeable reduction in Young's modulus. This particularly affects some polyamide types. The moisture that has penetrated causes the plastic to swell somewhat. Even though this is barely measurable, the molecular ball expands, reducing the attractive forces between the individual molecular chains. This causes Young's modulus to drop sharply.

Moisture absorption leads to a reduction in Young's modulus

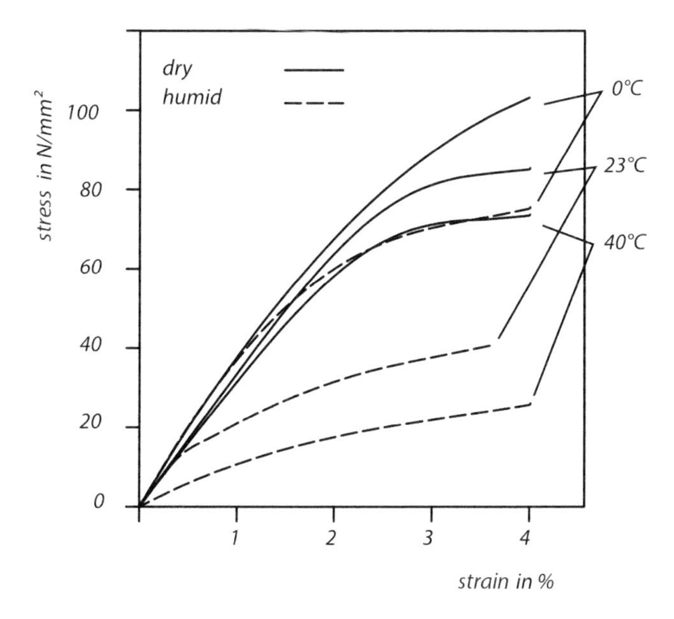

Figure 5.9 Influence of moisture on Young's modulus

When subjected to an impact load, a plastic will be considerably more brittle in its dried state than in its moist state. The glass transition temperature is thus reduced by moisture absorption, so if a material that is brittle overall absorbs moisture at room temperature, it loses its brittleness or it then only becomes brittle again at lower temperatures.

Moisture lowers the glass transition temperature

Polyamides, in particular, are greatly affected by moisture. Freshly molded components are always dry because the material must be dried before processing. After production, the component absorbs moisture again, depending on the ambient conditions; this is referred to as conditioning.

For metals, the tensile test is a static test; creep only occurs at high temperatures (approx. 0.3 T_m). Particularly in the case of semi-crystalline plastics, creep is not insignificant at normal service temperatures, including room temperature. Therefore, in addition to the tensile test, the creep test is essential.

Creep is a significant factor for plastics, even at room temperature

Test specimens are loaded with a constant stress for a very long period. The constant strain is noted and plotted in a diagram. In this way, pairs of results are ob-

Isochronous stress-strain diagrams

tained for one strain each at a given time and stress. These value pairs can be transferred to another diagram. Here, the corresponding stresses and strains are plotted at the same times (isochronous). Such diagrams are called isochronous stress-strain diagrams.

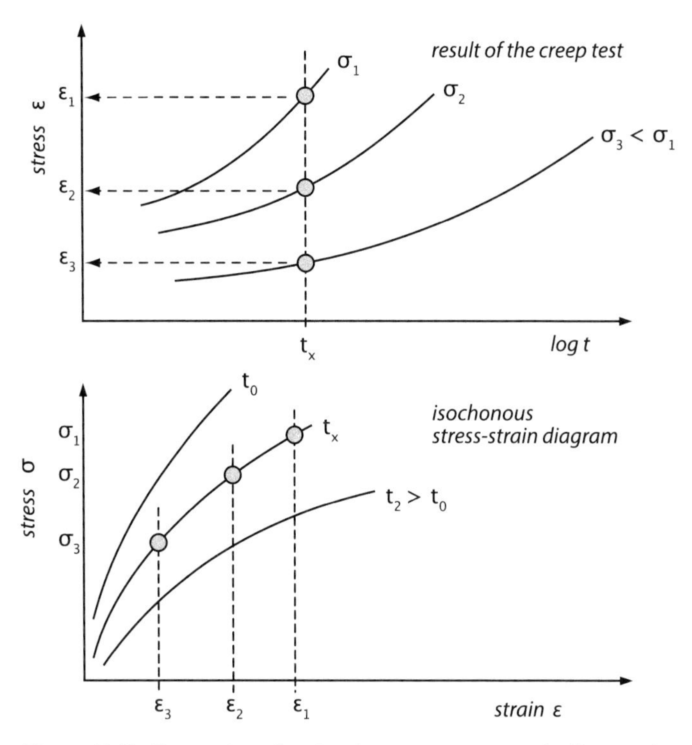

Figure 5.10 Generation of an isochronous stress-strain diagram from tensile test results

The use of isochronous stress-strain diagrams is very important for components subject to long-term loading. They can be used to read off the time after which a component will have permanently deformed under an expected permanent load.

Creep modulus from isochronous stress-strain diagrams

The creep modulus E_t is determined from the results of the creep test or from the isochronous stress-strain diagram.

$$E_t = \frac{\sigma}{\varepsilon(t)} = \frac{F}{A}\frac{L_0}{\Delta L(t)} \tag{5.6}$$

This is not the specification of the initial slope. The creep modulus is formed by the secant from the zero point and a stress value at a corresponding time deformation $\varepsilon(t)$. If the creep modulus is again plotted on a time axis, it becomes smaller and smaller with increase in load time.

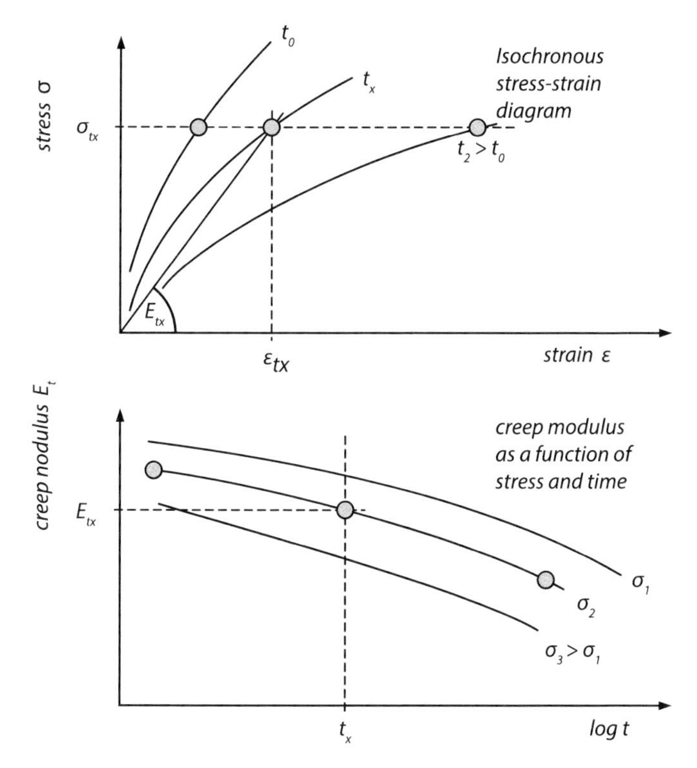

Figure 5.11 Forming the creep modulus from the isochronous stress-strain diagram

Creep modulus for dimensioning permanently loaded components

In engineering mechanics, Young's modulus is essential for dimensioning. It can be used to calculate the stiffness of a component or the extent of deformation under load. Creep modulus is used in a similar manner, but it must be remembered that the creep modulus is not a material constant. Particular attention must be paid to the stress time to be selected.

Problem of missing creep test data

Due to the high effort involved in the determination of creep modulus, isochronous stress-strain diagrams are available for only a few plastics, and so creep moduli are rarely available. In this case, reduction factors can be used to estimate the creep modulus using the short-term tensile test value. According to *Oberbach* (1975), the maximum permissible stress determined in this test is halved if a loading period of two years is assumed. This method is far from accurate, but a better method has still not been developed.

In any case, this method is appropriate, because the creep behavior is of course also temperature-dependent. Thus, the creep tests mentioned would have to be conducted at different temperatures.

5.2.5 Temperature Function of Young's Modulus

Dynamic mechanical analysis (DMA) provides information about the temperature-dependent behavior of a plastic

The mechanical behavior of plastics is strongly temperature-dependent. Of particular importance here is the dependence of Young's modulus on temperature. Measuring Young's modulus via a tensile test only returns a single value at one temperature and is therefore insufficient. The solution is dynamic mechanical analysis (DMA).

In DMA, a specimen is subjected to cyclical loading while the temperature is continuously raised. The load leads to deformation. The loading is often twisting (torsion), in which case the deformation would be shear (shear loading). Here, a cyclical tensile load is used for better understanding (Figure 5.12).

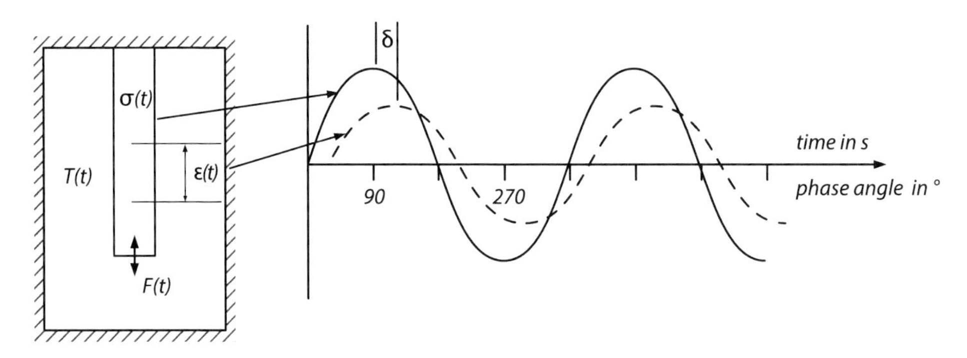

Figure 5.12 Principle behind dynamic mechanical analysis (DMA)

Time-delayed deformation in plastics compared to metals

In the case of an almost elastic material, e. g. a metal, the deformation occurs without delay. In a plastic, deformation can occur after a delay, especially at temperatures above the glass transition temperature. There, the material is viscoelastic, i. e. individual molecule groups can slide past each other.

Like electric currents, where a distinction is made between apparent power and reactive power as a function of the phase angle, a distinction is made between a storage modulus E' and a loss modulus E'' for an oscillation of the "response deformation ε" shifted by a phase angle δ. The storage modulus is directly comparable to Young's modulus. The two moduli are related via the loss factor "tan δ":

$$\tan\delta = \frac{E''}{E'} \qquad (5.7)$$

As a result of the oscillations at increasing temperatures, the storage modulus steadily declines. Above the glass transition temperature, the molecules can slide better and better, which is why the loss modulus increases here (Figure 5.13). With further temperature increase, the material can only absorb very little energy because the chains slide off immediately, and so the loss modulus decreases again. The loss factor tan δ is high for materials with a high proportion of non-elastic deformation.

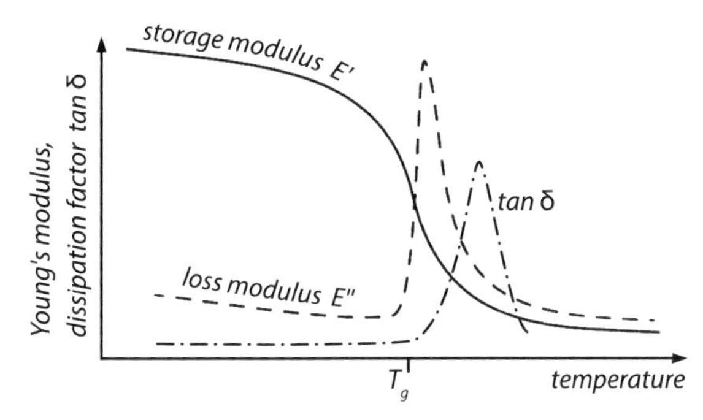

Figure 5.13 DMA evaluation: The storage modulus E' is largely Young's modulus as measured in the tensile test. The loss modulus E'' denotes the absorbed energy, tan δ is the ratio of E'/E''.

The loss modulus characterizes phase transitions of a material

For the user, the loss modulus is of no interest. However, a plot of the loss modulus provides important information about the material. Some plastics have a further maximum below the glass transition temperature; this is known as secondary softening (Figure 5.14). Smaller molecular groups, e. g. molecules at the side of the chain, become mobile.

Secondary softening is particularly important in polycarbonate (PC). Although PC, as an amorphous material with a glass transition temperature of 145 °C, should be brittle at room temperature, it is very tough at −100 °C due to secondary softening.

Storage module for dimensioning

The storage modulus can be calculated as a function of temperature with this test setup. It is directly comparable with Young's modulus as measured in the short-term tensile test. If Young's modulus is required for dimensioning, either the tensile test should be made at the temperatures relevant for the application, or the results from the DMA should be used direct.

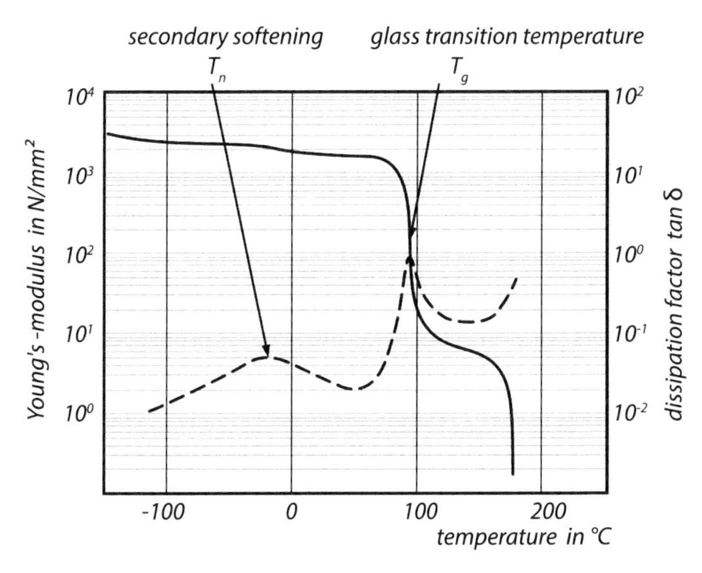

Figure 5.14 Young's modulus as a function of temperature of PVC

In any event, the temperature function of Young's modulus is very important for the designer. It shows the temperature ranges over which a strong change in mechanical behavior is to be expected. Possible failure cases can be better estimated, e.g. if a latching hook is to be actuated at low temperatures. Will it offer great resistance and possibly break or will it still be sufficiently tough and elastic?

5.3 Limits on Mechanical Design

Design limits for the dimensioning of plastic components: type of load, temperature, duration of load

For dimensioning, the load limits must be known. For metals, this is usually the yield strength $R_{p0,2}$ or the yield point R_e. At higher loads, permanent deformations will occur.

For plastics, the first questions to be clarified are the duration of the load and the temperature.

5.3.1 Short-Term Loads

The tensile test provides design information for short-term loads

The tensile test envisages three possible scenarios:

- *Brittle behavior:* The plastic breaks suddenly at load σ_B (breaking stress), without necking.
- *Tough behavior:* The plastic undergoes necking at load σ_Y (yield stress, Y stands for yield). Subsequent strain can be considerable.
- *Soft behavior:* The plastic deforms extensively even under low stress and the resulting curve is non-linear.

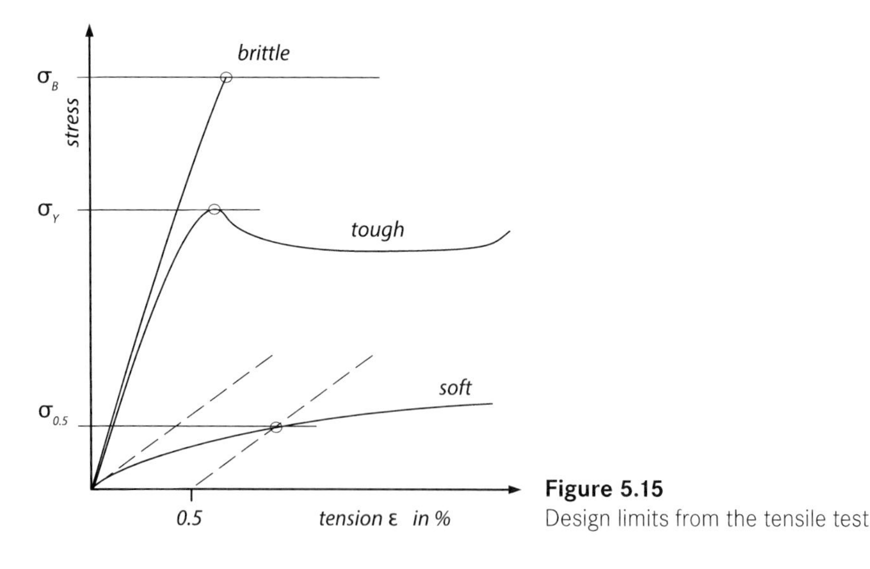

Figure 5.15
Design limits from the tensile test

The behaviors shown can occur in a plastic to an extent depending on the application temperature. This applies to semi-crystalline plastics above the glass transition temperature.

The design limits to be selected are accordingly the breaking stress σ_B, the yield stress σ_Y or the percentage stress $\sigma_{0.5}$. Of course, the 0.5% limit can also be shifted upward.

Design limits for components subjected to short-term loads

5.3.2 Long-Term Loads

For long-term loads, the results of the creep tests should be used first. Like metals in high-temperature service, permissible deformations are determined. These values can be used to determine permissible stresses, the exceeding of which would cause impermissible deformation.

Long-term loads can be estimated by means of creep tests

With amorphous plastics, creep behavior is not so pronounced. Here, a value of $\varepsilon_{F\infty} < 0.8\%$ should not be exceeded as a limit deformation. Particularly in the case of transparent components, permanent static load can lead to crazes.

Crazes are small, visible internal cracks

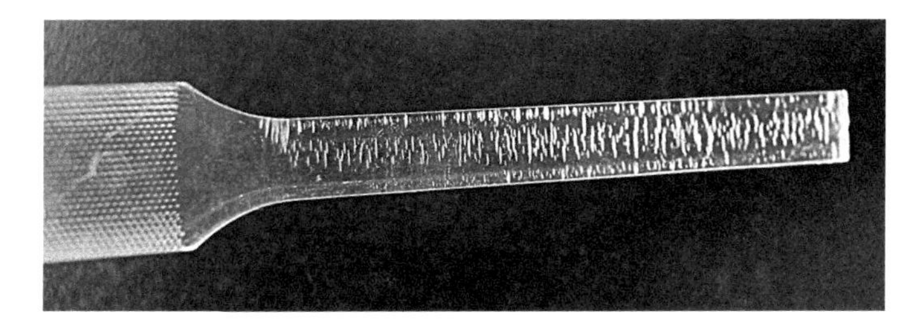

Figure 5.16 Crazes on a tensile test bar made of PS

Crazes appear as "micro-cracks". In fact, there are no cracks, because, when viewed under the microscope, many molecular bundles still bridge the crack. Under high strain, many small crazes are seen after a short time. With decrease in stress, the crazes appear later and are larger. Above a limit stress $\varepsilon_{F\infty}$, crazes no longer appear, even after a very long time. The strain is then approx. 0.8%.

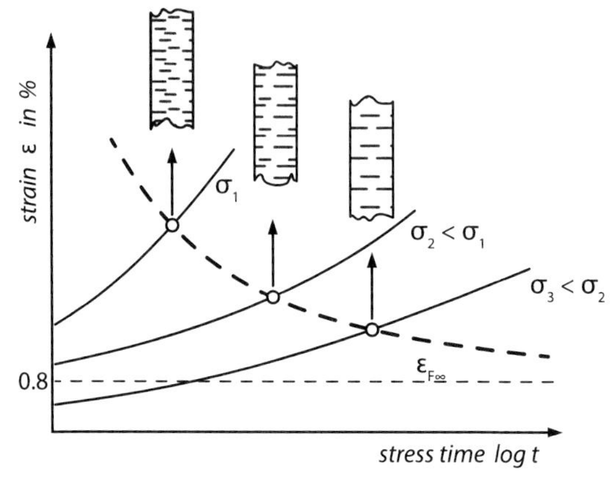

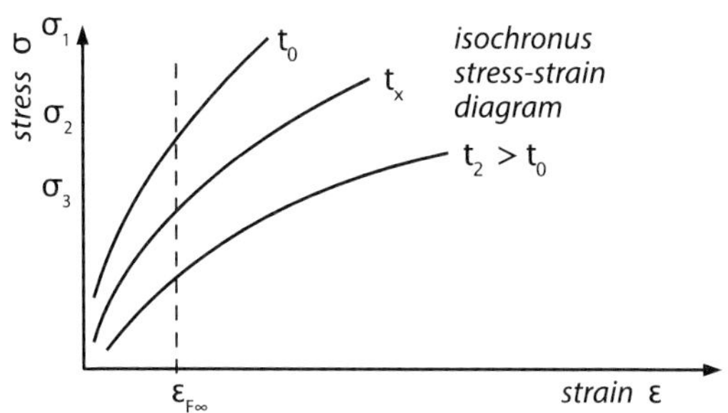

Figure 5.17
Critical strain for amorphous thermoplastics due to craze formation

Crazes, especially in the case of amorphous, transparent plastics

It is now advisable, especially in the case of transparent components, not to exceed a maximum strain of 0.8%. Then no visible damage in the form of crazes will occur.

5.3.3 Estimation of Design Limits Using Reduction Factors

Estimation of load limits

In many cases, time- or temperature-dependent data are unavailable. In this case, the desired limit values must be estimated from the available data yielded by the short-term tensile test. Here, the following applies:

$$\sigma_{allowable} < \frac{\sigma_{standard\ tensile\ test}}{A_{time} A_{temperature} S} \tag{5.8}$$

where S is the safety factor and A is a reduction factor. *Oberbach* recommends reduction factors specifically for temperature load and time duration:

- A_{time} – A few hours: 1.4 – Weeks: 1.7 – Months: 1.8 – Years: 2
- $A_{temperature}$:

$$A_{temperature} = \frac{1}{1 - k(T - 20)} \tag{5.9}$$

where T is the required temperature in k. Depending on the plastic, the following values can be used:

- ABS: 0.0117
- PA6: 0.0125
- PA66: 0.0112
- PA-GF: 0.0071
- PBT: 0.0095
- PC: 0.0095
- POM: 0.0082
- PP: 0.0116
- PE-HD: 0.0113

Index